Robert Scott Burn

The illustrated architectural, engineering, & mechanical drawing-book

For the use of schools, students, and artisans, upwards of 300 illustrations

Robert Scott Burn

The illustrated architectural, engineering, & mechanical drawing-book
For the use of schools, students, and artisans, upwards of 300 illustrations

ISBN/EAN: 9783742892713

Manufactured in Europe, USA, Canada, Australia, Japa

Cover: Foto ©berggeist007 / pixelio.de

Manufactured and distributed by brebook publishing software (www.brebook.com)

Robert Scott Burn

The illustrated architectural, engineering, & mechanical drawing-book

THE

ILLUSTRATED

ARCHITECTURAL, ENGINEERING, & MECHANICAL

DRAWING-BOOK.

FOR THE USE OF

Schools, Students, and Artisans.

UPWARDS OF 300 ILLUSTRATIONS.

BY ROBERT SCOTT BURN,

AUTHOR OF "THE ILLUSTRATED DRAWING-BOOK," "MECHANICS AND MECHANISM," "ORNAMENTAL AND ARCHITECTURAL DESIGN," "THE STEAM ENGINE, ITS HISTORY AND MECHANISM," ETC.

THIRTY-SECOND THOUSAND.

LONDON:
WARD, LOCK, AND CO.,
WARWICK HOUSE, SALISBURY SQUARE.

UNIFORM WITH THIS VOLUME.
Price 2s. each.

THE ILLUSTRATED DRAWING BOOK.
Comprising Pencil Drawing, Figure and Art Perspective, Engraving, &c. With 300 Illustrative Drawings and Diagrams.

THE STEAM ENGINE:
Its History and Mechanism. With 310 Illustrations.

MECHANICS AND MECHANISM.
Elementary Essays and Examples. With 250 Illustrations.

ORNAMENTAL DRAWING & ARCHITECTURAL DESIGN.
With Notes, Historical and Practical. With 300 Illustrations.

LONDON: WARD, LOCK, & CO., SALISBURY SQUARE, E.C.

THE
ILLLUSTRATED
Architectural, Engineering, and Mechanical
DRAWING-BOOK.

INTRODUCTION.

In the work on *Practical Geometry*, in the Series of Educational Books of which this treatise forms a part, we have given simple definitions and constructions of the various forms and figures which may be said to constitute the foundation of all drawing. We have there endeavoured to show that a knowledge of geometrical construction is necessary, before a thorough appreciation of the principles of outline sketching can be obtained, and a ready facility acquired in performing its operations. However much this position may be controverted as regards its application to an art which is generally looked upon as independent of, rather than dependent on, strict and formal rules, there can be no doubt, we think, that it holds with all completeness in reference to that which it is now our duty to illustrate and describe. In fact, so much do the various branches treated of in the following pages depend upon a knowledge of geometry, that many class them under the generic title of "geometrical drawing." Those commencing the study of these arts—so useful to the architect and the mechanic—without this knowledge of geometry, will be disappointed as to their speedy proficiency, and will labour under great disadvantages, from not understanding the principles upon which the constructions are founded.

In carrying out the objects of the present work, we purpose strictly to adhere to the plan followed in *The Illustrated Drawing-Book*, of adopting a series of progressive lessons, leading the pupil from the consideration of simple examples up to those more complicated in their construction. So that the simple steps are well understood, the more difficult ones will be easily mastered by the pupil who attends to the various gradations of examples. We have endeavoured, as far as the discursive nature of the subject has admitted, to preserve a distinct classification of objects, and a unity in the examples, so as to make the pupil thoroughly conversant with one department before proceeding to the consideration of another. Where this has been departed from, and an apparent mixing up of examples has resulted, considerations involving obvious advantages have suggested the change. On the whole, however, we trust that the classification so desirable has in some measure been obtained.

Although aware that architectural and mechanical drawing has for some time taken its place in many scholastic establishments as a branch of ordinary education, we are nevertheless anxious to see it still more extensively adopted. We conceive it likely to be of more general use—even to those who may not at all contemplate following up any of the professions to which it is more specially useful—than may at first sight be acknowledged. Apart from the habit of method, which, if it does not create, it will at least foster and encourage, we see many advantages accruing to those desirous of having a knowledge of science by an acquaintance with its practice. And there are few, we think, in these days of practical science, who are not likely to be interested in its progress. Geometrical drawing—taking the term in its widest sense—is an art which will enable those acquainted with its principles to understand a scientific exposition with greater readiness than those can do who are ignorant of it. To convince the reader of the truth of this, we have only to remind him, that few expositions of improvements or inventions in practical science, in its widest range, are ever made without the aid of sketches,—these ranging from the simple diagram up to the more complicated drawing; and the ready understanding of these is open only to those acquainted with drawing. This consideration should, we think, weigh very forcibly with those who are doubtful of the propriety of following the example of so many educational establishments, in introducing geometrical drawing as an ordinary branch of education. To those desirous of following out the profession of architect, engineer, or mechanic, an acquaintance with the art is as indispensable as a knowledge of sketching from nature or objects is to the artist or painter. Without it, the practical man, however ingenious, will inevitably fail in perfecting, unaided, his ideas with that facility available to the accomplished draughtsman. Its usefulness in the workshop, moreover, is no less conspicuous than in the study or bureau, in enabling the inventor or improver to communicate his ideas clearly and readily to the workman. To the latter also it is equally important and indispensable,—we mean to those who are desirous of raising themselves above the level of the mere operative, the handler of the hammer or the mallet. In short, to him who, in the exercise of his important avocations as architect or engineer, wishes to render himself independent of extraneous assistance in planning and working out his original ideas, and capable of communicating their results to others with facility, a knowledge of the art is absolutely indispensable. To those who have at all considered the subject, further comment on its value to the practical man is altogether unnecessary.

One cause probably of the art not being so generally adopted in educational establishments, is the extreme paucity of books treating exclusively on the subject; and of those calculated to serve as guides, the price is nearly prohibitive, at least to the generality of purchasers. A work taking up all the departments of the subject, treating them *methodically* and *fully*, yet issued at a price absolutely within the reach of all, has long been a desideratum; which we venture to hope the present volume may possess some claim to having supplied. As to its *method*, we have followed the same plan which, adopted in the *Illustrated Drawing-Book*, met with a considerable amount of success. If we have failed to attain in this the same clearness of exposition and attractiveness of illustration, this may be attributed, in some degree, to the nature of its subjects, which are not in themselves so attractive as those contained in the above work. At all events, we may lay claim to a strong desire to make it, in all its departments, as attractive and useful as possible. As respects the *fulness* or completeness of the work, a mere glance at its pages will show that, even should fault be found with the method of the lessons, none can be alleged on the score of the meagreness of their illustrations. The present volume is strictly a "drawing-book," showing how drawings—whether architectural, engineering, or mechanical—may be copied and laid down. The æsthetic principles and technical rules which *dictate* the proportions, the style, and the methods of planning structures—whether these be architectural or mechanical—are not treated upon, excepting in some few instances. These rules and principles, we conceive, belong to more strictly professional treatises. A work which is contemplated in this series—on "Ornamental and Architectural Design"—will, in one department, provide a guide to the pupil, desirous of going into the principles of one art, of which we here give lessons useful in delineating its examples. A work giving a clear exposition of the principles and practice of the other departments, namely, " Civil and Mechanical Engineering," is yet a desideratum among educational books. Should it be thought advisable, such a treatise may probably be included in the present series.

R. S. B.

PREFACE TO THE SECOND EDITION.

THE Author feels gratified that the demand for another Edition of this Work has enabled him, by a careful revisal of every example, to render the description by which it is elucidated as clear and explicit as possible, and to expunge many errors and crudities which encumbered the first issue. Various improvements in the plan of the book have been suggested to him, some of which he has endeavoured to embody in the present issue. The majority of these, however, are more intimately connected with the department of a work to which frequent reference is made in this volume, namely, that on " Ornamental and Architectural Design," and the utility of which is adverted to in the Introduction above given. This work was suggested to the Author during the preparation of the present volume, some years ago; and since that period he has had numerous evidences afforded him of its probable utility in an educational course. From these the work is considered a desideratum, which it is deemed desirable to supply. For this purpose, the Publishers of the present volume have kindly entrusted its preparation to the Author; and it is hoped that in a short time it will be ready for publication. The treatise is intended to embody a variety of examples in the different styles of *Ornamentation*, and of the methods by which they are drawn; together with notes, historical, æsthetical, and practical, on *Architectural Design;* the whole having an immediate aim to their application to the general purposes of utility, as well as being calculated to afford a variety of useful information to the general reader.

R. S. B.

THE
ILLUSTRATED
Architectural, Engineering, and Mechanical
DRAWING BOOK.

ARCHITECTURAL DRAWING.

In this department the first lessons will be those which require for their construction nothing but the arrangement and combination of right or straight lines. It is scarcely necessary here to state, that the instruments requisite for the various operations are the same as those required for the constructions described in the *Illustrated Practical Geometry*, the 'drawing-board,' 'square,' and 'triangle,' these being absolutely indispensable. As this work is strictly designed to be a sequel to that on Geometry, we beg to refer the reader thereto for a description of these, and the readiest methods of using them.

EXAMPLE 1. *To draw the portion of 'hipped roof' shown in fig. 1.*

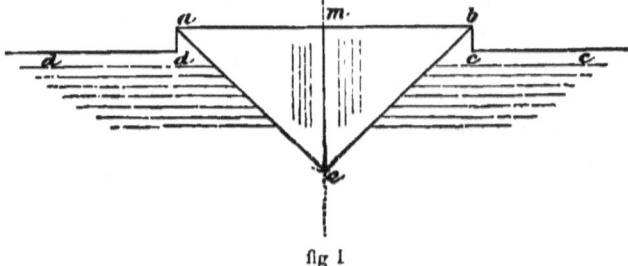

fig 1

On the drawing-paper, properly fastened to the board, mark any point a; parallel to the side of the drawing-board, draw from this point any line, and make it equal to ab. Bisect this in the point m; at right angles to ab draw from this point a line to e, and make this equal to me in the copy. Join ea, eb. At right angles to ab, from these points draw to cd, making the length of these equal to bc in the copy. Draw the lines cc, dd, parallel to ab. The example given in fig. 1 will thus be copied. The pupil should be very careful to take the measurements in his 'compasses,' or 'dividers,' equal to those in the copy.

EXAMPLE 2. *To draw the part plan of the wall of a house, showing the projection of one side of fireplace, with the internal flue, in fig. 2.* Divide the thickness of the wall $g'g$ by the centre line ab, bisect fe, and draw at right angles to ab a line $cm\,dn\,n'$. These lines are to be drawn on the copy in *light pencil lines*, as also the line g produced to d and h. These preliminary operations being performed on the paper on the drawing-board on which the copy is to be made, draw any lines ab, cn, corresponding to those made on the copy in fig. 2. From the point c of intersection, with the measurement cn' taken from copy, lay off from c to n'; with distance $n'f$, from copy, lay off on a line drawn through n' at right angles to cn', to the points e and f; from e draw a line at right angles to ef to the point g, and make it equal to eg; do the same at f, and make the line equal to fv. Lines drawn from g and v, parallel to ab, will represent the internal line of wall; the external, or external line $g\,1$, will be put in by measuring its distance on the copy from the centre line ab, and transferring it to the paper on the board, and thus drawing the line $g\,1$ parallel to ab. The next portion to be copied is the internal flue. Take the measurement dn from the copy, and transfer it to the corresponding line on the board; in like manner put the measurement nm; with no or ms, on lines drawn at right angles to cn', measure to s, o, and join the points so, so; the example is completed. Another method of copying this figure may be adopted as follows:—Assume on the paper on which the drawing is to be made any point e fig. 2; parallel to the side of the board draw a line ge, and make it equal to the line ge in the copy. At right angles to this draw ef, and make it equal to ef; from f, parallel to eg, draw fv; from g and v draw lines, as in the copy, parallel to fe; parallel to these, and at the proper distance, put the line $g\,1$. With the distance nn' from e lay on the line eg, and through this point draw a line ono, parallel to fe. At the point m, the distance of which from the point n is easily obtained from the copy, draw another line sms; from the line ge measure to the line so, and transfer it to the board; parallel to ge draw so, meeting the lines ms, no. Measure the breadth of the flue from o to o, transfer it to the board, and join the upper ends of the lines ms, on, by a line so. We have here shown two methods, chiefly to enable the pupil, by a ready exercise of his reasoning powers, to decide as to the quickest method of copying any figure presented to him.

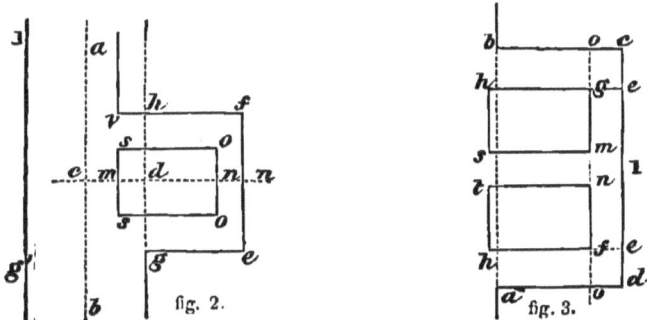

EXAMPLE 3. *To draw one jamb with two internal flues.* Let ab, fig. 3, be the internal line of wall, $abcd$ the outline of jamb. Produce the sides hg, hf, to meet the line cd, in the points ee, and mg, nf, to meet bc, ad at the points o, o. Having fastened the paper properly on the board, and

proceeded with the copy as directed above, the produced lines being marked in light pencil lines, the first operation is to draw any line, as *a b*, on the most convenient part of the paper on the board on which the drawing is to be copied. From *a* measure to *b*; and from these points, at right angles to *a b*, draw lines to *c* and *d*; make *b c*, *a d*, equal to the corresponding lines in the copy; join *d c*. Next take from the copy the measurement from *d* to *e* (the point found by producing *h f*), and lay it from the point *d* in the board on the line *d c*. Do the same from *c* to *e*; parallel to *b c* draw lines *e h*, *e h*; measure next the distance from *d* to *o*, and transfer it on *a d*, *b c*, to *o o*; parallel to *a b* draw a line *o o*. From *g* and *f* measure to *h h*; transfer these, and from the points obtained draw a line *hh* parallel to *o o*. With distance *h s*, or *n f*, measure from *h* to *s*, from *g* to *m*, from *n* to *f*, and from *t* to *h*; join the points. In inking the lines, the points *b c*, *d a*, *h g*, *s m*, *t n*, *h f*, will be their terminations. In the examples given in the figures the lines not dotted show the complete design.

EXAMPLE 4. *To draw the outlines of an ordinary sash window.* Make any line *c d* on the paper on the board equal to the corresponding line in the copy, fig. 4. At the points *c d* draw lines perpendicular to *c d* of an indefinite length. With the measurement *c a*, from the copy, cut the lines *c a*, *d b*, in *a*, *b*; join *a b*. Divide the line *c d* into three equal parts in the points *n*, *n*; parallel to *a c*, from these draw lines meeting *a b*. Divide the line *a c* into four equal parts in the points *m*, *e*, *g*; parallel to *a b*, from these draw lines meeting *b d* in *m'*, *f*, *h*. The parallelogram *a b d c* is divided into twelve lesser ones, representing each a pane of glass.

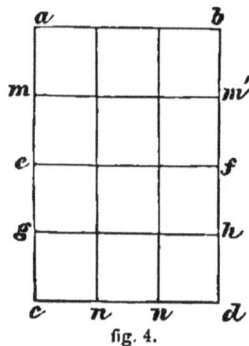
fig. 4.

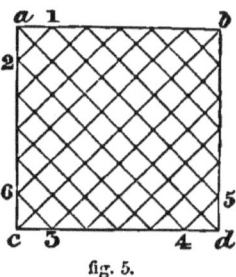
fig. 5.

EXAMPLE 5. *To draw the diagonal lines representing the panes in a rustic window, fig. 5.* Draw *c d*, making it equal to *c d* in the copy; *c d* is the side of the square *a b d c*, which describe. Divide the sides *c d*, *a c*, each into six equal parts; join the corresponding points, as 1, 2, 4, 5, 3, 6.

EXAMPLE 6. *To draw the central pilaster in fig. 6.* Divide in the copy the line *a b* into two equal parts at *c*, and through this, at right angles to *a b*, draw the line *c c*; bisect the part *o o* by a line *s f s*. On the paper on which the drawing is to be made, draw any line representing *c c*, and another at right angles to the first, representing *a b* in the copy, fig. 6. The intersection of these two lines will give the position of the point *c*. Take from the copy the measurement *c f*; transfer it from *c* to *f*; draw parallel to *a b* through this point a line, representing the line *s f s* in the copy. With *a c* from c lay off on ab to a, b; draw indefinite lines from these points perpendicular to ab; measure ad, be equal to ad in the copy; join ed. With measurement ft, lay off from f to t and u, and on the line cc as many times as necessary; through these several points draw lines parallel to ab; these will be the centre lines of the parts corresponding to oo. With af from the points of intersection of these with the line cc, lay off equal to oo; through the points draw lines parallel to ab; produce ab to gg. The terminations of the parts equal to oo will thus be formed, as represented in fig. 6. In the copy, as in fig. 4, produce the internal lines hh, to meet the line ab in mn; from c, with cm lay off on ab to m, n, and parallel to cc from these points, draw lines mh, nh. These lines will terminate the alternate internal portions. Another method of copying this figure will be as follows:—Draw any line ab, and at right angles to it another, bg; the point where they meet will correspond to the point b in the copy, and thus a datum point will be obtained from which to take measurements. With ba from the copy, set off ab parallel to de; from a draw dg; make ad, be each equal to the corresponding lines in the copy; join de. From e measure to the line above it, and transfer it to the paper on the board; from the same point measure to the next line; and so on in succession. Transfer these measurements to the corresponding points on the paper on the board, and through the points obtained draw lines parallel to ab; these will form the under and upper lines of the parts oo. The lines representing the boundary-lines of the alternate inner portions can be obtained by measuring from e or d to the lines as mh, nh, in the copy, and transferring them; thereafter through the points obtained drawing lines at the parts required parallel to bg.

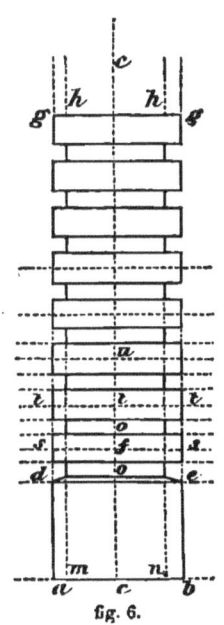

fig. 6.

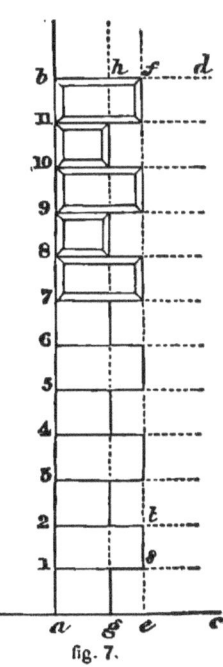

fig. 7.

EXAMPLE 7. *To draw the quoins of a house in fig.* 7. Produce in the copy the external line *f* to meet the base line *a g* produced in *e*; next, on the paper on the board draw any lines *a c*, *a b*, at right angles to each other; then the point of intersection will correspond to the point *a* in the copy. Measure the distance *a g* from the copy, and transfer it to the board; do the same with *a e*. From these points draw lines parallel to *a b*, as *g h, f e*. The line *a b* will represent the corner line of house, the line *g h* the internal line of quoins, and *e f* the external. Suppose *a b* to be the height on which the quoins are to be disposed, make *a b* on the board equal to *a b* in the copy; and on the supposition that there are to be twelve quoins in *a b*, divide *a b* into twelve equal portions, and through the points thus obtained draw lines parallel to *a c*, as *b h*, *f d*, 1 *s*, 2 *t*. Finish as in the copy. Another method is as follows:—Draw lines *a c*, *a b* as before; measure from *a* to *g*, and draw from this a line parallel to *a b*; measure from *a* to 1, and draw through the point a line 1 *s*; measure 1 *s*, transfer it to the paper on the board; measure *s t*, and draw it at right angles to *a e*; join *t* 2 by a line parallel to *a c*. The first quoin 1 *s*, *t* 2 will then be drawn, and afford datum points from which to finish the others; thus the line *s t* produced towards *f* will give the external line of all the others; and the distance *a* 1, transferred in succession to the line *a b*, will mark the horizontal distances.

EXAMPLE 8. *To draw the figure in fig.* 8, which represents the plan of the roof of an outbuilding, or external addition, projecting from the main wall *f n s*. The dotted line *c d* must be first drawn in the copy, dividing

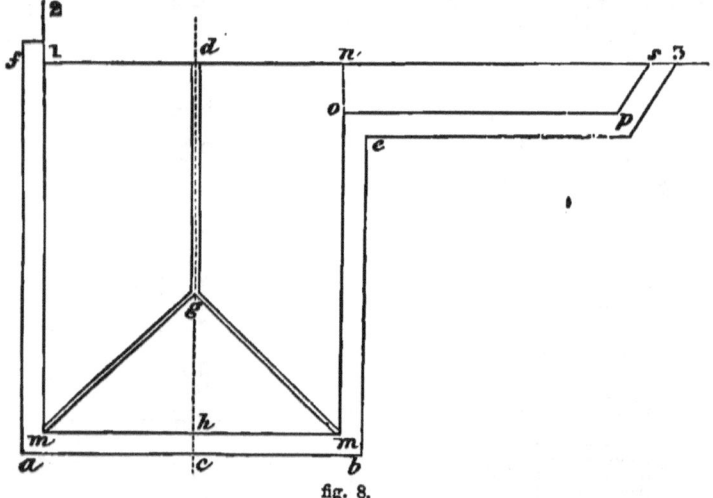

fig. 8.

a b into two parts at *c*; next, on the paper on the board draw any two lines at right angles corresponding to those in the copy, as *c d*, *f d n s*. To avoid unnecessary repetition, we wish the reader to understand that, when we give directions to measure any part or distance, as "measure from *c* to *b* and *a*," we mean that the distance *c b* is first to be taken from the copy in the figure, and transferred to the corresponding point on the board—thus ascertaining the position of the points b, a corresponding to those in the copy. The copy, as in the figures, is, in all instances, the only source from which measurements are to be taken: nothing in this species of drawing is to be left to the eye,—all must be tested by the instruments. Inaccuracy of measurement in any one point will inevitably result in throwing the whole drawing wrong. Thus, for instance, supposing the distance $a\,1$, in fig. 7, was taken with the smallest possible error in measurement—say too much—it would be found that the distance would not go twelve times between ab, but would go beyond b to a much greater distance than would be supposed. Where an erroneous measurement is to be transferred from one point to another in succession, the original error increases in a remarkably quick ratio. But to proceed with the consideration of the construction of figure 8. Measure from c to b and a, and draw $acbd$; from ab parallel to cd draw lines to f and e; measure from c to h; through h draw a line parallel to ab; from h measure to m,m. Or these points may be obtained by measuring from the lines af, be. From m, m draw lines parallel to cd to $1\,n$; from n measure to o, and parallel to $1\,3$ draw a line to p. Measure from n to s; join sp. From c measure to g; draw the lines gd, and join gm, gm. A line from 3 parallel to sp, joining a line from e parallel to op, will complete the figure.

EXAMPLE 9. *To draw the plan of part of roof in fig. 9.* Bisect in the copy the line between cc in the point b, and draw ab. On the board draw any line ab corresponding to ab in the copy, and, at right angles to

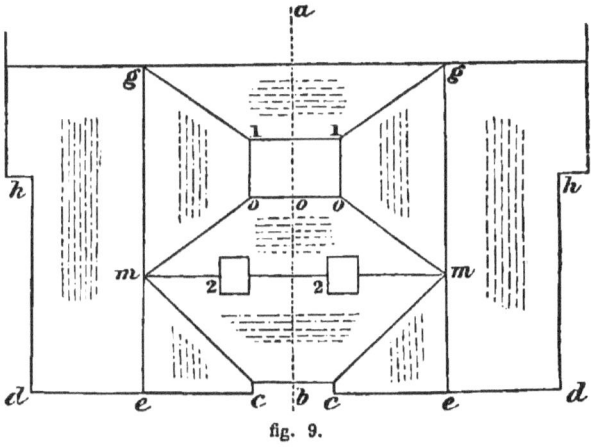

fig. 9.

it, another, representing the lines cd, de, cd. At c,c drop the perpendiculars, as in the copy, and join them by a line parallel to cd. From c measure to d, and parallel to ab draw lines dh, dh; measure from d to h. From c measure to e, and parallel to ab draw eg, eg; measure from e to g, and draw a line gg parallel to cd. On eg, eg, measure to mm; draw mm parallel to cd. From b measure to o, and put in the lines oo, $1\,1$; join $o\,1$, $o\,1$: this represents the cistern in the roof for rain water. Join $g\,1$, $g\,1$, mo, mo, these representing the sloping lines of roof. From m measure to 2, and put in the plan of chimney flues.—(*See figs.* 1, 2.)

EXAMPLE 10. *To draw the window in fig.* 10. Bisect in the copy ab in c, and draw a line cf; draw lines on the board corresponding to ab, cf. From c measure to a and b, and draw lines ae, be. From c measure to n, d, m, s, h, and f, and through all these points draw lines parallel to ab,—that drawn through d deciding the length of the lines ee. From n measure to nn, a distance equal to half nn in the copy. Do the same at the points f and m to mm, gg. Parallel to fc from n, n draw to i, i, and join hi, hi. From g, g parallel to cf draw to m, m. From d lay off to oo, and from these draw lines to mm, parallel to cf, meeting the line mm; join these with the point s by lines parallel to hi, hi. From i measure to k', and make $k'k$ at right angles to oi. At right angles to $k'k$ make kp; from p measure to r, and draw rt parallel to $k'k$. Measure tx, and put in the line 11.

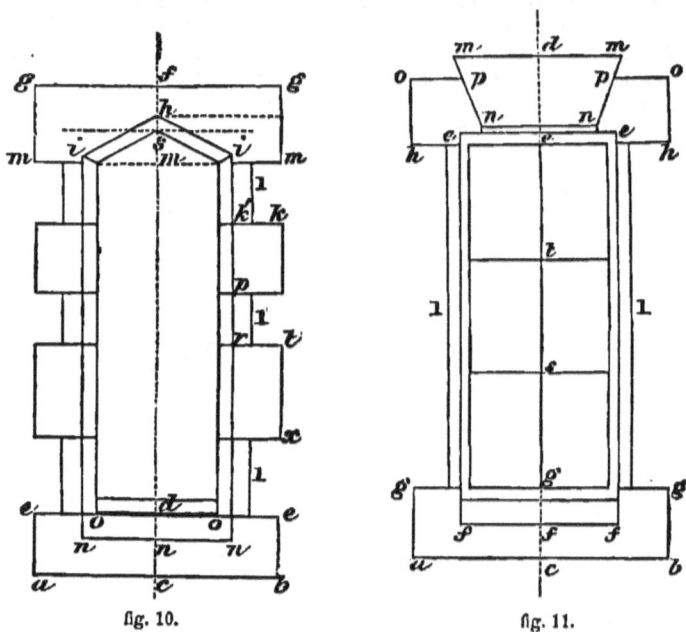

fig. 10.   fig. 11.

EXAMPLE 11. *To draw the window in fig.* 11. Bisect ab in c in the copy, and draw cd. On the board draw any two lines corresponding to ab, cd in the copy; measure from c to a and b, and draw from these points to gg lines parallel to cd; measure to gg, and join ggg. From c measure to e and d and f; through these draw lines parallel to ab. From f measure to ff, and from e to ee; join fe, fe by lines parallel to cd. From e measure to h, h and to n, n; from d measure to m, m, and join mn, mn. From hh parallel to cd draw lines to oo; measure ho, and parallel to hh draw lines from oo meeting mn, mn. Draw the parallelogram within ee, ff, and from c measure to s and t, and through these draw lines parallel to ab; these represent the divisions of the glass. Put in the lines 11 parallel to cd, joining gg, hh: the drawing is complete.

EXAMPLE 12. *To draw the form of window in fig.* 12. Draw the centre-line cc, as in preceding examples. Corresponding to the lines ab, cc draw two on the board. From *c* measure to *a b*, and from the same point to *f*, *m*, and *c*; through *c* draw a line parallel to *a b*, as *e e*, and through *f*, *f f*, and *m*, *g g*. Measure from *f* to *f*, *f*, and from *m* to *g*, *g*; join *g f*, *g f*. From *c* measure to *x*, and join *g x*, *g x*. Parallel to *a b* put in the lines *d d*, *h v*, *h v*. From *v*, *h*, parallel to *g f*, draw lines to meet *g*, *g*, and from the points *o*, *o* lines to meet these, parallel to *g x*. From *e* measure to *n*, and put in *n t*, *n t*; put in the square *s s s s* by lines parallel to *m o*, *g x*. Draw the external 'dressings' by the method described in fig. 10.

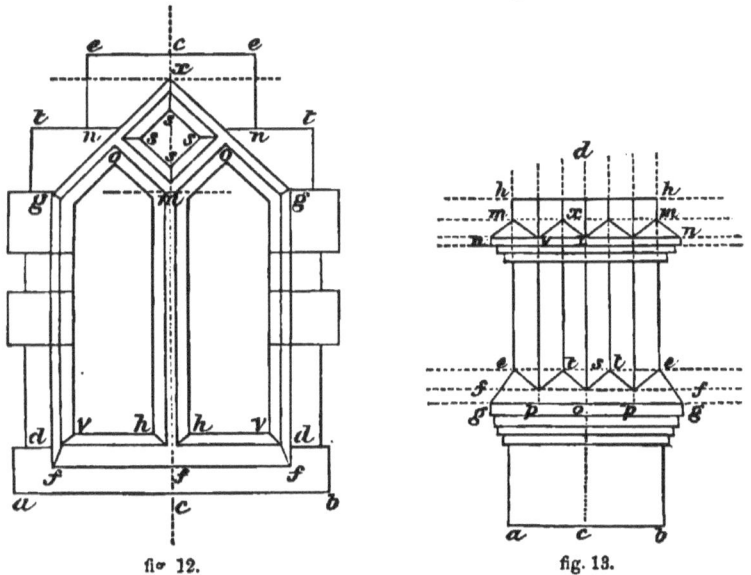

fig. 12.   fig. 13.

EXAMPLE 13. *To draw the chimney-shaft in fig.* 13. Bisect the line *a b*, draw a line *d o c* through this; draw corresponding lines to *a b*, *c d* on the board; measure from *c* to *o*, *s*, *x*, and *h h*; measure from *o* to *g g*, and put in the part *g g*, as well as those receding parts under it. Through *s* draw a dotted or *occult* line* *e s e*, divide *o g*, *o g* into two parts at *p p*, with the half of *o p*, from *s* lay off four times to *e e*, on both sides of the line *c d*, join *g e*, and from *t*, *t* lines meeting *f f*. Produce the lines *e p t* and *s* beyond the line *h h*; measure from 1 to *n*, *n*, join *n m*, *n m*, *v x*, and *x* 1, and put in the remaining portion as by preceding lessons.

EXAMPLE 14. *To draw the steps of a staircase as in fig.* 14. Let *d g* be the height from one line of floor to the other, represented by the upper and under lines; and *c d* the distance in which the steps are to fall. The height of each step is 7 inches, technically called a 'riser;' the breadth being usually 9 inches; this part on which the foot rests is called the 'tread.' The measurements in the figure are taken from a scale one-fourth of an inch to the foot. Suppose the width *c d* to be 6 feet, allowing 9 inches for the tread, this will give eight divisions; divide *c d* therefore into eight equal parts, and from these points draw lines perpendicular to *c d*. Taking the

* For definition of the various kinds of lines see *Illustrated Practical Geometry*.

height dg, from one landing to another, to be 5 feet 3 inches, this will give nine divisions of 7 inches each; divide dg, therefore, into nine equal parts, and from the points thus obtained draw lines parallel to cd. From the intersection of the line 1 on cd, and 1 on ab at a, draw to the point of intersection of 2 the line from 1, with that from 2 on dg; from the point n, the intersection of the lines 2 2, draw a line meeting the intersection of the vertical line 2 with the horizontal 3. The intersection of the

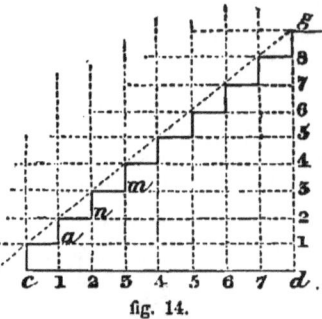

fig. 14.

lines 3 3 gives the point of next step, and so on, each time proceeding nearer the line ag.

EXAMPLE 15. *To delineate the plan of the stairs in the preceding lesson.* The distance ab, fig. 15, corresponds to dg in fig. 14; the breadth ab being that between the side walls or balustrades. The height ab is divided into 9 equal parts, each part representing a step. If a line be drawn from the point 9, fig. 14, to the left-hand top corner of the front step at c, it will be found to touch the corners of all the steps; this forms the foundation for another method of delineating the profile of the steps in a staircase, as described in

EXAMPLE 16. Let 1 2, fig. 16, be the breadth, and 2 b the height from one landing to another, as before; draw the line 1 3, and join b 3. From b on the perpendicular b 2, make off to c a distance equal to the height of one

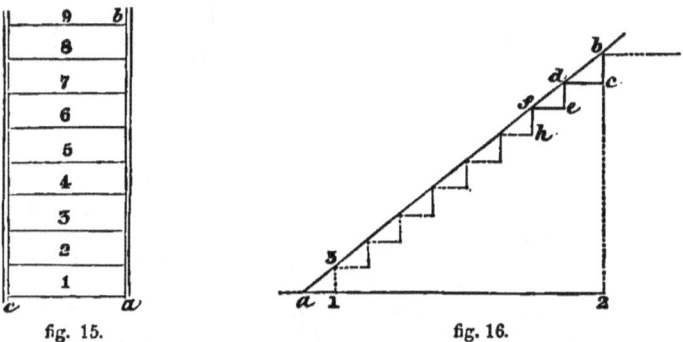

fig. 15.   fig. 16.

riser equal to a 1 3. From c draw a line exactly parallel to a 2, or perpendicular to b 2, meeting the diagonal line b 3 in d; from d drop a perpendicular de, equal 7 inches, or bc; from e draw parallel to cd a line meeting b 3 as before; from f drop a perpendicular to h, and proceed thus till finished. Great care must be taken to draw the lines truly parallel to the proper lines; also to drop the perpendiculars, as de, exactly from the point where the horizontal lines, as cd, join the diagonal b 3. The least deviation from accuracy in the beginning will inevitably result in throwing the operations towards the end far wrong. The lines should be drawn very finely, so that the exact points of intersection will be easily observable. The

B

method shown in fig. 14 will be least liable to error. We give the two methods, as affording opportunities of extended practice to the pupil, and as suggestive of plans he may himself adopt.

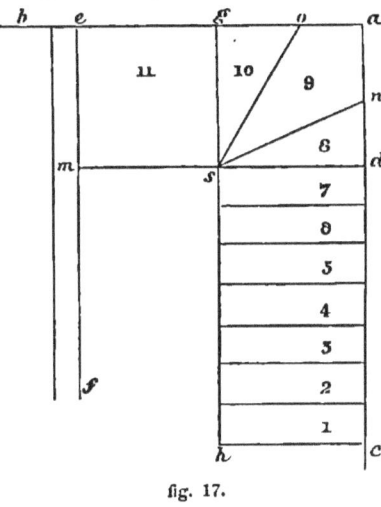

fig. 17.

EXAMPLE 17. *To delineate the plan of a staircase having 'returns,' by which the direction of the steps is changed.* Assume any point in fig. 17, as *a*; draw *a b*; perpendicular to it draw *a c*. Measure from *a* to *e* and *g*, draw from these points *e f*, *g h* parallel to *a c*; from *a* measure to *d*, and draw *d m* parallel to *a b*; measure from *d* to *c*, and draw the line 1; divide *d c* into seven equal parts. From *d* measure to *n*, and from *a* to *o*; join *s n*, *s o*. We give another lesson similar to this in

EXAMPLE 18. Draw *a b*, *b d* at right angles to one another; from *b* measure to *c* and *e*, and draw *e h*, *c h* at right angles to *b d*, *b a*. Divide *c d* into two equal parts, and *c a* into seven. Measure from *c* to *f*, and from *b* to *g*; join *h f*, *h g*.

EXAMPLE 19 shows plan of cellar-steps having a return at head which is entered from *p*, and another at foot entered from *s*. A party-wall is between the two houses, the steps of the adjoining house being shown in dotted lines.

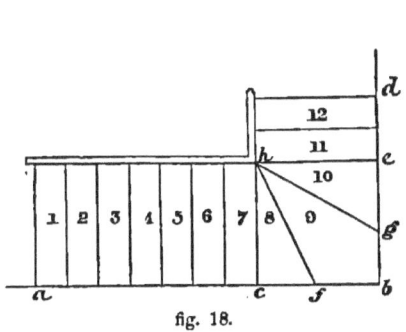

fig. 18.

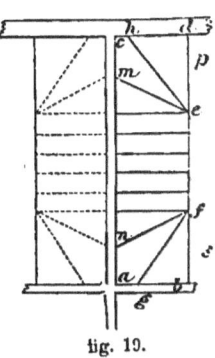

fig. 19.

Draw *a c*, *a b* at right angles; from *a* measure to *c*; draw *c d*, representing the inside line of external wall, parallel to *a b*. From *a* and *c* measure off the thickness of party-wall, dividing the two staircases, and draw a line *d b* parallel to *a c*. From *d* and *b* measure to *e* and *f*; from these points draw lines parallel to *a b*. From *c* measure to *m*, and from *a* to *n*; join *e m*, *n f*. Measure from *d* to *h*, and from *b* to *g*; join *f g*, *e h*. Divide the distance between *e f* into as many equal parts as in the drawing; from these points draw lines parallel to *a b*; these represent the steps in the stair parallel to the party-wall.

ENGINEERING, AND MECHANICAL DRAWING-BOOK. 19

EXAMPLE 20. In fig. 20 we give a sectional vertical sketch, showing a flight of stairs c, reaching to the first landing-place d', from the ground-floor $a\,b$, with return steps e, leading to the first-floor f; the landing-place d, counts as one step, a step d', rising into the room of which $d'd$ is the door; $a\,a$ is the door of a room on the ground-floor, $g\,g$ of one on the first-floor. The ground-plan of this is shown in

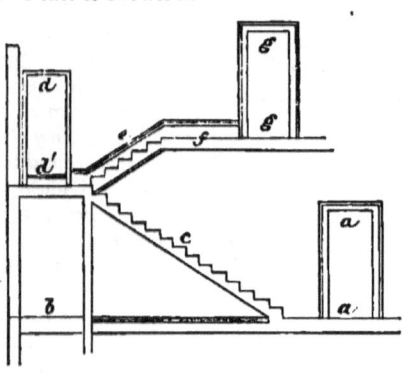

fig. 20.

EXAMPLE 21, fig. 21, the letters of reference in which correspond with those of fig. 20. The first flight, $c\,c$, is drawn in full lines, the return being dotted. The chamber or first-floor plan is shown in

EXAMPLE 22, fig. 22. The steps of the first flight are shown in full to where the landing reaches and the banisters begin; the dotted lines represent the steps, which are hid by the flooring-boards of the chamber floor.

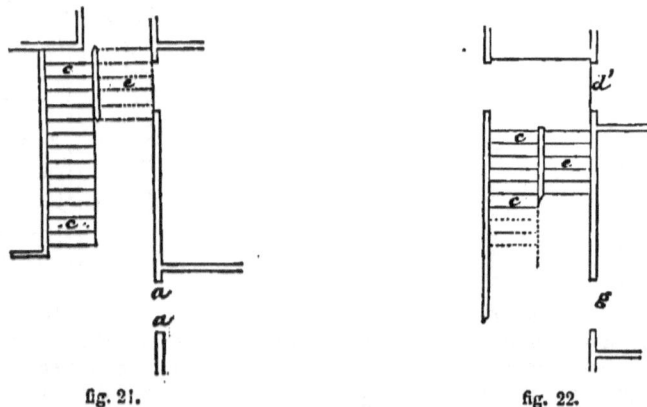

fig. 21.   fig. 22.

We now proceed to the consideration of lessons in which a combination of straight and circular lines are met with. The first lesson is given in

EXAMPLE 23, fig. 23. Let $a\,b$ be the ground-line of a house, and with part of the circular-headed window of apartment in basement seen above it; divide the width into two equal parts at c, draw $c\,d$ perpendicular to $a\,b$.

B 2

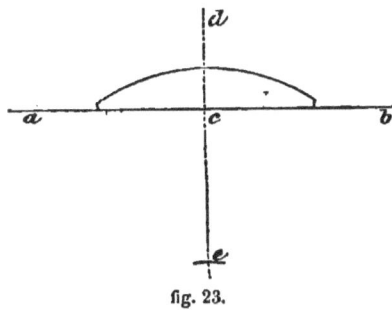

fig. 23.

Draw lines on the board corresponding to ab, cd; the centre of the curve will be found somewhere on the line de. By trial on the copy, it will be found to be at e; measure from c to e, and from c as centre, with cd as radius, draw the curve, joining the short perpendicular lines from ab. The centre from which the part of the circle in the copy is described can easily be found by adopting any of the methods described in *Practical Geometry*, one of which is here shown in fig. 25.

EXAMPLE 24. *To draw the circular-headed fire-place in fig.* 24. Let ab be the width, bisect it in c, draw cd perpendicular to ab; draw corresponding lines to ab, cd on the board; measure from c to a, b, from these draw lines perpendicular to ab, and equal to af. From c as a centre, with cf as radius, describe the arch fde. Measure from a and b to h, h to get the width of jambs, and perpendicular to ab draw hm, hm; measure to m, and draw mn parallel to ab; measure to o, and parallel to cd draw lines oo, oo.

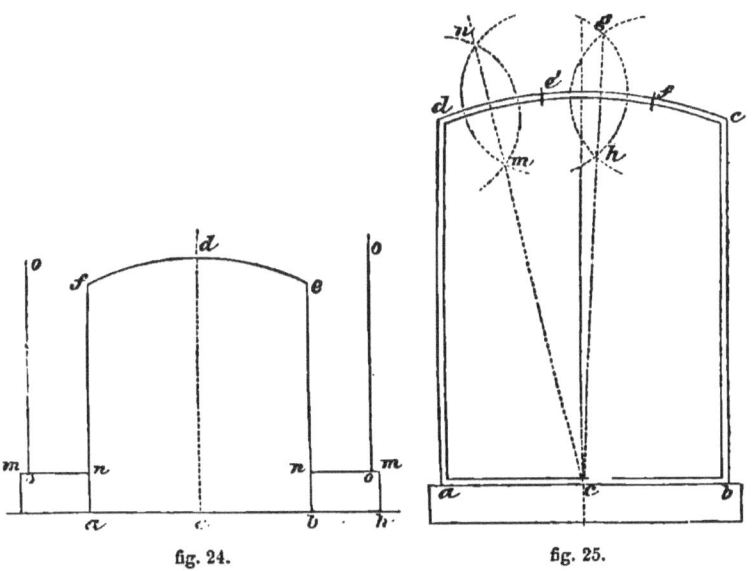

fig. 24. fig. 25.

EXAMPLE 25. *To draw the circular-headed window in fig.* 25. Let ab be the width, bisect it in c, draw cg perpendicular to ab. From a, b parallel to cg draw lines to d and e, the termination of the curve. To find the centre from which the curve dfe is described, take any points e' and f in the copy, and from these points, with radius greater than half the distance between them, describe arcs cutting in g, h. From d and e' in like manner describe arcs cutting in *n m*; through *n m*, *g h* draw lines meeting in *c*; *c* is the centre from which the curve is described. The centre may be found also by trial on the line *c d*, as described in fig. 23. The sketch may be copied by transferring the various points found, to the paper on the board, proceeding as in the foregoing lessons.

EXAMPLE 26. *To draw part of cellar-plan of house in fig. 26, showing walls, top of 'copper,' and flue of furnace connected therewith.* The sketch *without any of the dotted lines* is supposed to be given to copy from. By trial in the copy find the centre of the circle, which will be at *o*; from *o* draw a perpendicular to *a b*, parallel to *a s*, and another line at right angles to this, as *o e*, touching the line *c a* in the point *e*. On the paper on the board draw any two lines intersecting each other at right angles, the point of intersection at *o* will represent the point *o* of the copy. Measure from *o* to *d*, and from *o* to *e*: draw at these points lines at right angles, meeting in the point *a*. From *a* measure to *b*, draw *b f* parallel to *d o*; in the copy

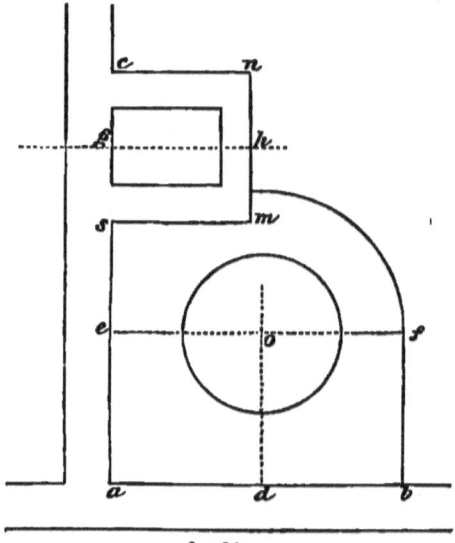

fig. 26.

bisect the side *n m* of the flue, and draw the line *g h* at right angles to *m n*. From *h* measure to *m*, *n*, and from these points draw lines meeting *c a* in *c*, *s*. From *o*, with proper radius, describe the circle, and from same point with *o f* describe part of a circle, joining *f* with side of flue *m n*. Another method of copying this may be adopted. Draw any two lines *c a*, *a b* at right angles, meeting in *a*; from *a* measure to *s*, at right angles draw from this point a line and measure *s m*; from *m* parallel to *c a* draw a line *m n*; from *n* parallel to *s m* draw a line meeting *c a* in *c*. The internal flue can be put in, as shown in fig. 2. From *a* measure to *b*, draw *b f*. Find by trial the centre of the circle, measure the distance of this from the two sides, *a s*, *a b*, transfer these to the board, and describe the circle as before.

EXAMPLE 27. *To draw the walls and cellar-flues given in fig.* 27. In the copy continue the line *d* across *a*, the line *g q* across *f*, *t t* across *m*. By trial find the centres of the circles *p, p*, join them by a line *p o p*. On the board draw any line *a b*, representing the centre line of the wall *f m*, and at right angles to it another *d d*. From *a* measure to *e* and *c, c*, from these draw lines forming a parallelogram, as in the copy. From *a* measure to *f*, and through *f* draw a line parallel to *d d*; from *f* measure to *g g*, put in *h, h*, and join *g, h* by lines perpendicular to *g g*. From *a* measure to *o*; draw a line

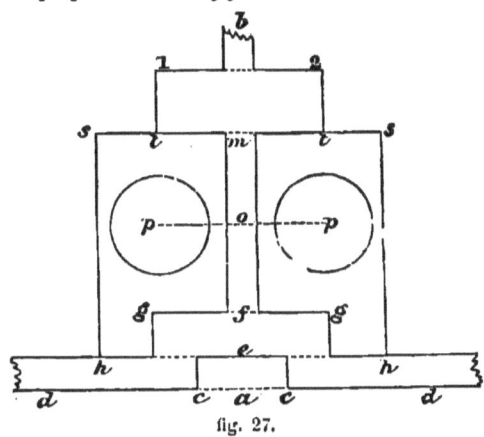

fig. 27.

through this parallel to *d d*; from *o* measure to *p p*; these points are the centres of the circles. From *a* measure to *m*; draw a line as before, and measure to *s, s*. From *m* measure to *t, t*, and draw lines to 1, 2; put in the thickness of the wall *m f*.

EXAMPLE 28. *To draw the 'bull's eye' in fig.* 28. Bisect *e e* in *d, f f* in *d*, and from these draw lines intersecting at *a*. On the board draw lines corresponding to these. From *a*, with *a h* as radius, describe the circle *a h*. From *a* measure on the four radial lines to *d d*, from *d* measure to *e, e*, &c.; join *e d e*. From the points where the circle *a h* cuts these lines, measure from *h* to *m*; join *f m*; do this at all the radial lines. From *a*, with *a s* as radius, describe the parts of a circle joining the key-stones, as *t t*, &c. From *a*, with *a h*, describe in like manner a circle, as *m o*. From *a* measure to *c, c*, and from *c* to *b*; from *a*, with radius *a b*, describe parts of a circle, joining both ends of the lines *c, c*; finish as in the sketch.

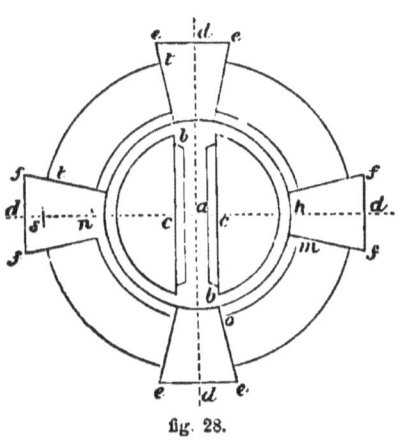

fig. 28.

ENGINEERING, AND MECHANICAL DRAWING-BOOK. 23

EXAMPLE 29. *To draw the bracket of a cornice in fig.* 29. Let *a a* be the line of wall; from *a* draw *a b* perpendicular to *a a*. Measure from *b* to *c*, and draw *b c*. Measure from *a* to the point 5, and draw from it a short line

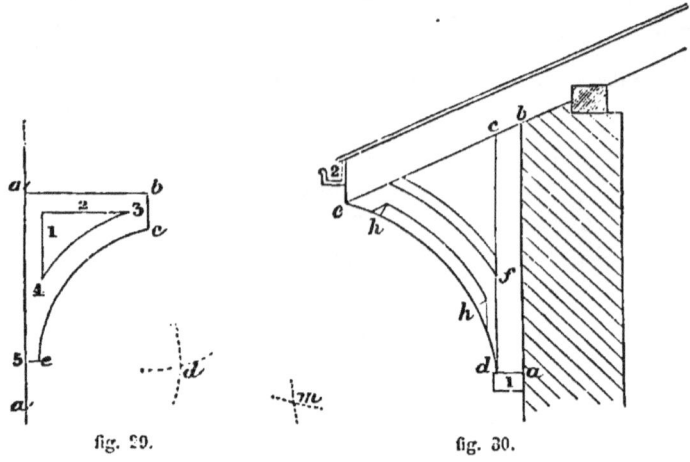

fig. 29.   fig. 30.

perpendicular to *a a*, as in the diagram. From *a b* measure to 2, and from *a a* to 1; from these draw indefinite lines. Measure from the line *b c* to the point 3. By trial find the centre of the circle *c e* in the copy, as *d*, and transfer it to the board; join by an arc *e e*.

EXAMPLE 30. *To draw another form of bracket-cornice.* Draw a line *a b*, put in the part *a d* 1. Draw *c d* parallel to *a b*. From *a* measure to *b*, and from *d* to *c*. By drawing a line exactly through these points, the angle of the line of roof *e c b* will be obtained. From *c* measure to *e*; by trial find in the copy the centre from which the curve *e d* is described, as *m*; with the radius thus found, from the points *d* and *e* on the board, describe arcs cutting in *m*; with same distance still in the compasses, from this point describe a curve joining *d e*. From *d* and *e* measure to *h h*, and put the other arcs in as shown.

EXAMPLE 31. *To draw the window in fig.* 31. Draw two lines, *a b*, *c d* at right angles to one another, intersecting in the point *c*. Measure from *c* to *a* and *b*, and also to *d*; through *d* draw a line *e d e* parallel to *a b*; measure from *d* to *e*, *e*; join *a e*, *b e*. Measure from *c*

fig. 31.

to *m*, and draw through this a line parallel to *a b* ; measure also to *n*, and draw *n n*. From *m* measure on both sides the distance *m o* ; *m n* also from *n* to *n, n*; these points are the centres of the circles shown in the sketch, the method of putting in of which is still further elucidated by

EXAMPLE 32. Let the line *c d*, fig. 32, correspond to *o m n* in fig. 31, *a b* to *n m*, and *f e* to *n n*. From the point of intersection of these lines with *c d*, describe the circles as in the drawing. On each side of *c d* draw the

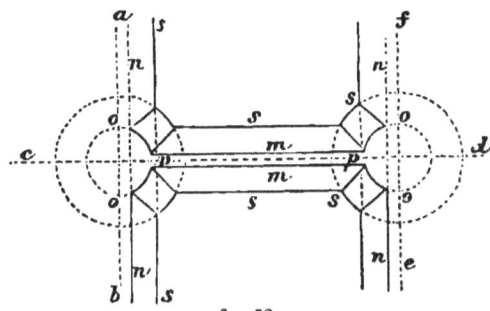

fig. 32.

lines *m, m* ; and parallel to same lines, lines *s, s* touching the circles. From *f e, a b*, lay off to *n n* lines equal to *m* ; and from *m*, to *s s* equal to the distance of the line *m s* from *c d* ; join *p s* and the points corresponding.

EXAMPLE 33. *To draw the basement arch below the principal entrance to a house, as in fig. 33.* Draw the line *a b*, and at right angles to it a line from *c*, the centre of *a b*. Measure from *c* to *a* and *b*. From same point,

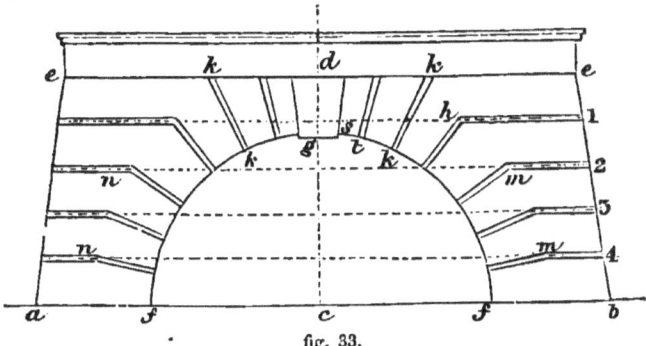

fig. 33.

with *c f* as radius, describe the semicircle *f f*. From *c* measure to *d*, draw a line through this parallel to *a b*. Measure from *d* to *e, e*; join *a e, b e*; put in the key-stone *d g*. Divide *b e* into five equal parts, and from these points, parallel to *a b*, draw lines through *d c* to the line *a e*. From *s* measure to *t*, and draw lines on each side the key-stone *d g*, parallel to its sides. From *t* measure to *k*. Divide *k f* into five equal parts. From *i* measure to *h* ; from *c*, with *c h*, describe a dotted semicircle passing through the points *n n*, *h m m* ; this will give the termination of the lines drawn from the points on *b e*. Join these with lines to the points found in the part of the circle *k f*.

EXAMPLE 34. *To describe the ornament (part of a verandah) in fig. 34.* Let ab be the breadth; bisect it in c, draw cd at right angles to ab. Draw on the board lines corresponding to these; the line cd will be that on which the centres of the complete circles are found. From c measure to a and b; draw af, be; the centres of the parts of circles *within the complete ones* will be found on these lines. At any distance on ab draw a line gmh parallel

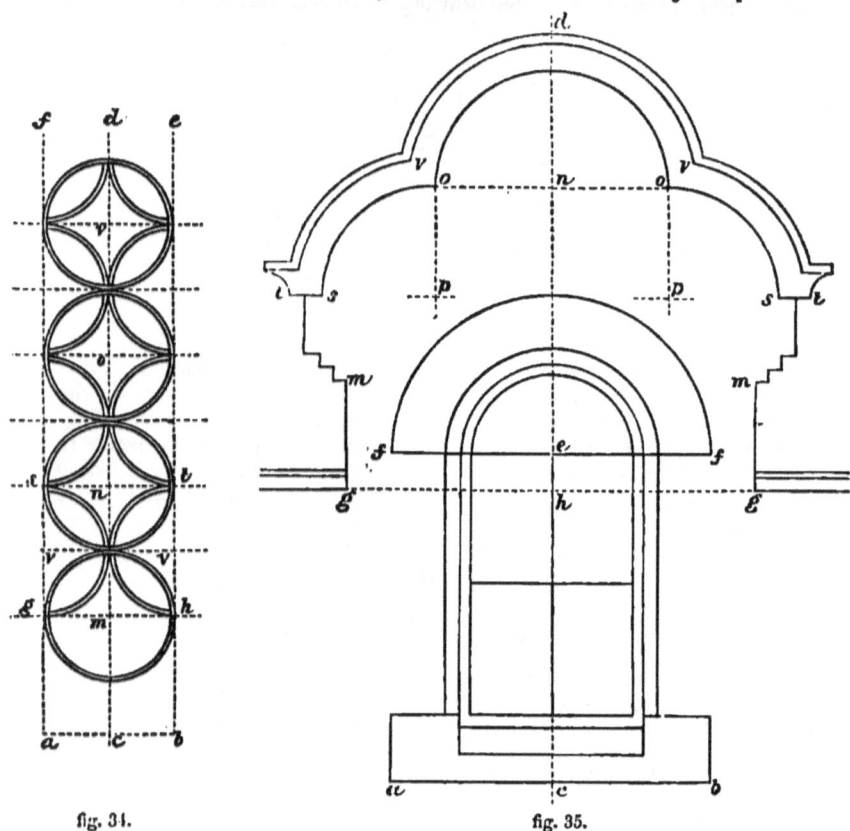

fig. 34.   fig. 35.

to ab. With ac, from the point m, describe a circle gmh. With gh, the diameter of the outer circle, lay off on cd from the point m to the points n and o. Through these draw lines parallel to ab, as snt. From n, with radius ac, describe a circle snt. Through the point where the two circles touch, draw a line vv parallel to ab, cutting af, be. With radius a, c, from v, v, describe semicircles as in the sketch. The centres of the remaining circles will easily be found from the foregoing instructions.

EXAMPLE 35. *To draw the window in fig. 35.* Bisect ab in c; draw cd; join gg and oo by dotted lines as in the copy. On the board draw lines corresponding to ab, cd. From c measure to a, b, and put in the cill acb, as described in fig. 10. From c measure to h, e, and n. From h measure to

$g\,g$, and from these points draw lines parallel to $c\,d$; draw $g\,m$, $g\,m$. From c with cf describe the semicircle; and from n, with $n\,o$, $o\,n\,o$. Perpendicular to $o\,n\,o$ draw lines to p, p; with the radius of the circle $o\,n\,o$ measure to p, p; from these points with same radius describe the quadrants $o\,s$, $o\,s$. From s draw $s\,t$ parallel to $a\,b$. Finish the circles as in the copy. The method of putting in the part from g to v will be more fully described in

EXAMPLE 36. Let m, p in fig. 36 represent similar points in fig. 35, $s\,o$ the inner circle, and $s\,t$ the horizontal line at termination of dripstone. From the point m draw $a\,m$ parallel to $t\,s\,p$; at a draw $a\,b$ equal and perpendicular to $a\,m$; from b, $b\,c$; from c, $c\,d$; and from d, $d\,c$; all equal to $a\,m$, and at right angles to one another. Join c to $t\,s$ by a line parallel to $p\,n$, as ef. Let $g\,o$ be the distance of the circle $g\,h$ from $s\,o$; from p, with $p\,g$, describe a quadrant to h, making the point h distant from the line $s\,t$ equal to $g\,o$. In like manner describe $n\,r$. From h and r draw lines $h\,m$, $r\,x$ parallel to $s\,t$. From the points $m't$, with radius greater than half the distance, describe arcs meeting in v; from v, with same radius, describe the arc $m\,t$; join $x\,m$.

EXAMPLE 37. *To draw the Elizabethan gable in fig.* 37. Divide $a\,b$ in the point c; draw $c\,d$. Corresponding to these draw lines on the board. From c measure to a, and put in the part below, as in the sketch. From

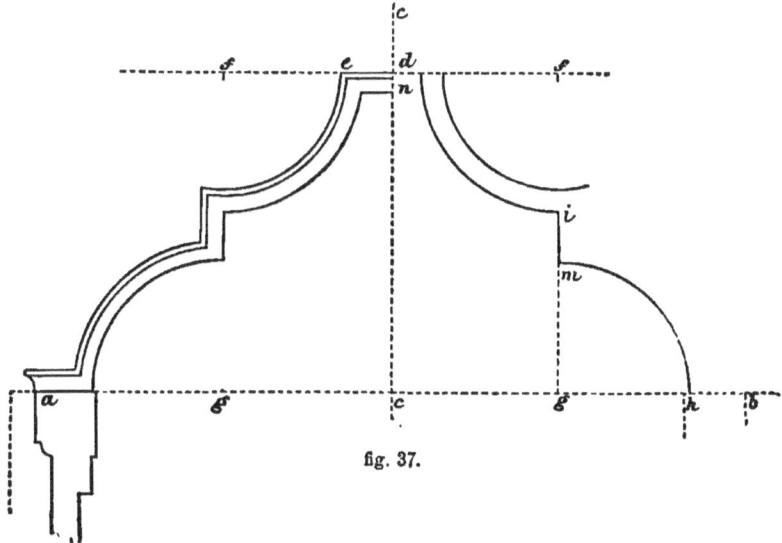

fig. 37.

measure to *d*, and draw *f d f* parallel to *a b*. From *d* measure to *e e* and *f f*; from *e* measure to *g*, and from *g*, with *g h*, describe the quadrant *h m*. From *m* draw *m i* parallel to *c d*; from *f*, with radius *f n*, describe the arc meeting the line *i*. Finish as in the part to the left of the sketch.

EXAMPLE 38. *To describe the flutes and fillets in fig.* 38. Let *ab* be the diameter of column, bisect it in *c*; draw *c d*. Draw on the board lines corresponding to these, and from the point *c*, with *c b*, describe the semicircle *a d b*, representing half of the column. Bisect the quadrant *a d* in the point

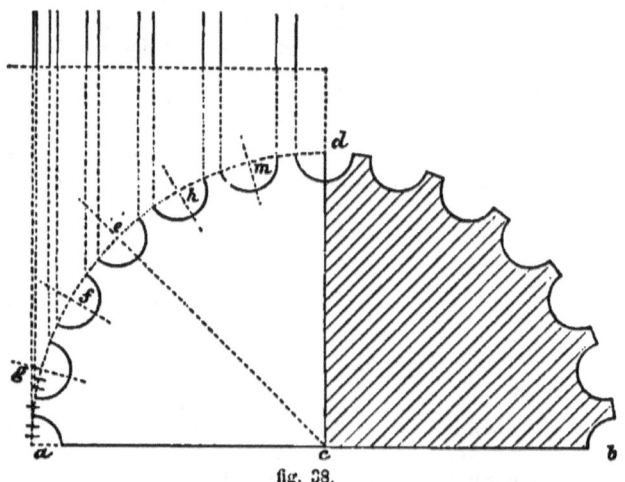

fig. 38.

e, and divide the arcs *a e*, *e d*, by points *g, f, h, m*. Mark the position of these by radial lines from *c*, as in the copy. Divide the part *a g* into eight equal parts; and with three of these as radius, from the points in the quadrant, as *g,f*, &c., describe semicircles. Six parts will thus be given to each flute, and two to each fillet: and the column will have twenty-four flutes.

EXAMPLE 39. *To describe the flutes in a Doric column without the fillets, as in fig.* 39. Proceed, as in last example, by dividing the quadrant *b e c* into

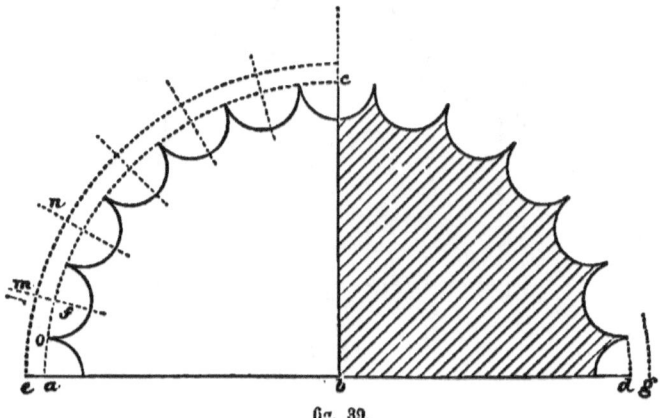

fig. 39.

six equal parts, as e, m, n, &c., giving to the entire column twenty-four flutes as before. Draw radial lines from b. Divide af into four equal parts, and lay one of these on ab produced to e; from b, with be, describe a semicircle as emn, cutting the radial lines. Bisect af in o, and with fo as radius, from the points—where the dotted semicircle intersects the radial lines—as centres, describe the arcs as in the copy. Another method is shown in

EXAMPLE 40, fig. 40. Describe a semicircle ade, and divide the quadrant bad into five equal parts, so as to give twenty flutes to the column. Produce ab to f; bisect ae in h, and from e lay off eh to m; join hm, and

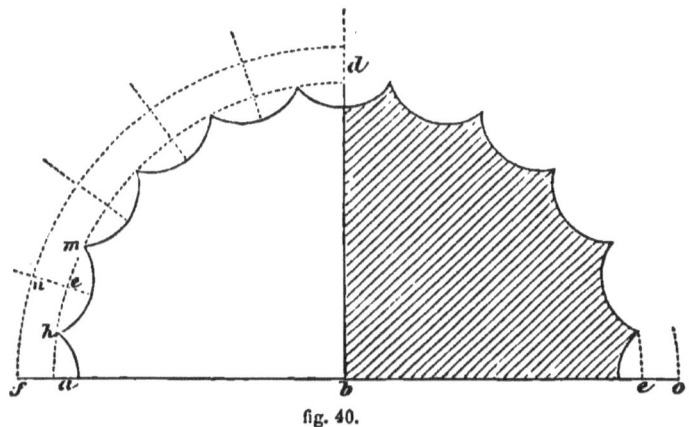

fig. 40.

with distance he lay off on the radial line be to n. From b, with bn, describe the dotted semicircle, fno. The centres of the flutes are placed where the radial lines intersect this semicircle. From n, with nm, describe the curve mh, and draw the others in the same manner.

EXAMPLE 41. *To describe the flat flutes and fillets as in fig.* 41. Describe the semicircle adc, and divide the quadrant bad into six equal parts;

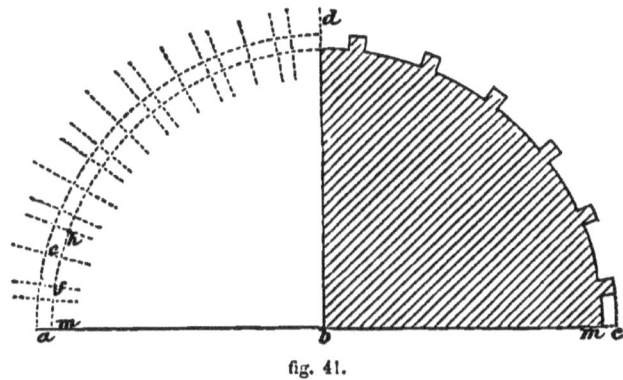

fig. 41.

divide ae into five equal parts. With two of these from the radial line, lay off on each side, as f, h. With one part lay off from c to m, and from b, with $b\,m$, describe a semicircle $c\,d\,a$; complete the diagram as shown. This will give the depth of the flutes, one; the width, four; and the width of fillets, one.

EXAMPLE 42. *To describe the cabled moulding in fig.* 42. Divide the semicircle $a\,c\,d$ in the same proportion as in fig. 38, giving an equal number

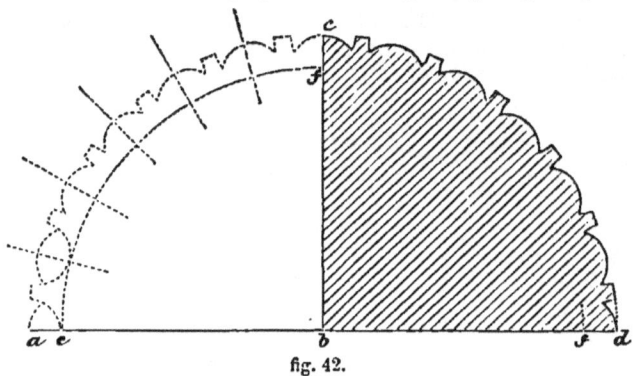

fig. 42.

as in that example. From b, with $b\,e$, describe the semicircle $e\,f\,f$. From the points where the radial lines intersect this, as centres, with radius $a\,e$, describe the curves as in the copy.

EXAMPLE 43. *To delineate the flutes in a pilaster,* fig. 43. Let $a\,b$ be the breadth; divide it into twenty-nine equal parts: each flute is three parts in breadth, and each fillet one. This gives to the pilaster seven flutes and

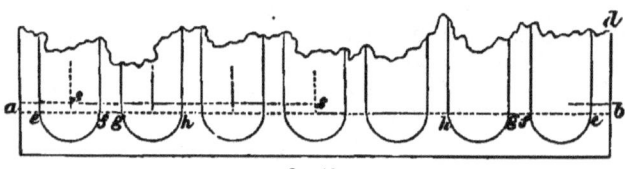

fig. 43.

eight fillets. Draw $a\,c$, $b\,d$ at right angles to $a\,b$; and parallel to these lines, from the first point next these, as at e; at the fifth of these points, as at f; the sixth, at g, draw lines. The first fillet is $a\,e$, the first flute, $e\,f$; $f\,g$ the second fillet, $g\,h$ the second flute, and so on. The centres from which the termination to the flutes are described will be on the line $s\,s$, this being intersected by lines drawn parallel to $a\,e$, drawn through a point bisecting the fillet $e\,f$, $g\,h$, &c.

EXAMPLE 44. *To describe the curves in the twisted Doric column in fig.* 44. Proceed as in

EXAMPLE 45, fig. 45. Draw the centre-line $a\,b$, and the line of base $c\,c$, the width $d\,d$ being that below astragal in capital; join $d\,c$, $d\,c$. With distance $c\,c$, lay off on $a\,b$ from a to e, and draw through this point the line $g\,h$, parallel to $c\,a\,c$. With half $c\,c$, as $a\,c$, lay off on $a\,b$ to f. From f as centre, with $f\,g$ as radius, describe the arc $g\,c$; with $f\,h$ as radius, from the points c and h as centres, describe arcs cutting in m; from m as centre, with $m\,h$ as radius, describe the arc $h\,c$. Make $e\,n$ equal $g\,h$; with $e\,g$, or $e\,h$, lay

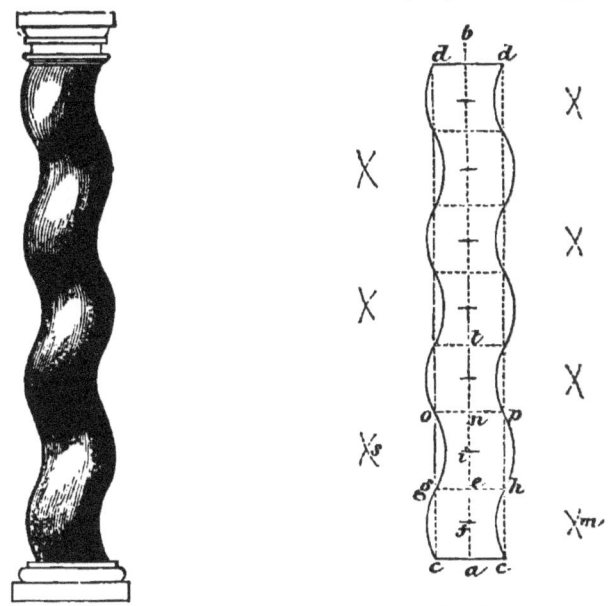

fig. 44.   fig 45.

off to i. From i, with $i\,p$ as radius, describe the arc $p\,h$; from the points g, o, with same radius, describe arcs cutting in s; from s, with same radius, describe the arc $o\,g$. Next make $n\,t$ equal to $o\,p$, and proceed as already described. The various centres are shown by the intersection of the arcs.

We now proceed to describe the method of laying out complete plans of houses. The first example of which we give in

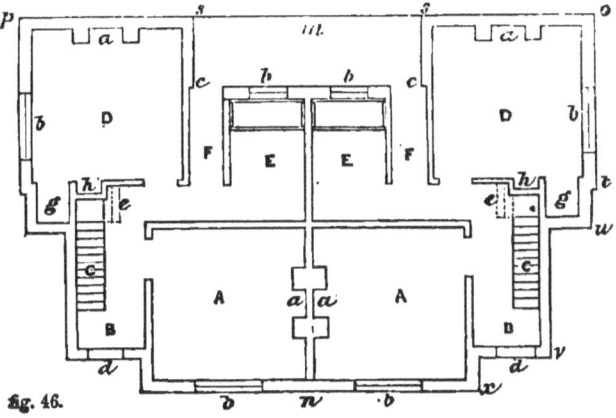

fig. 46.

ENGINEERING, AND MECHANICAL DRAWING-BOOK. 31

EXAMPLE 46, fig. 46, which is the 'ground-plan' of a pair of cottages, the division or party-wall being at *m n*, A A the living-rooms, D the kitchen, E the scullery, F the back lobby, B the front lobby: *a a* are fire-places, *b* windows, *d* doors. The method of copying this is given in

EXAMPLE 47, fig. 47. Draw the line *o p*, fig. 46, and bisect it, drawing from the point of bisection another line *m n* at right angles to *o p*; next, as in fig. 47, draw the lines *c d*, *a b* at right angles, corresponding to *p o*, *m n* in

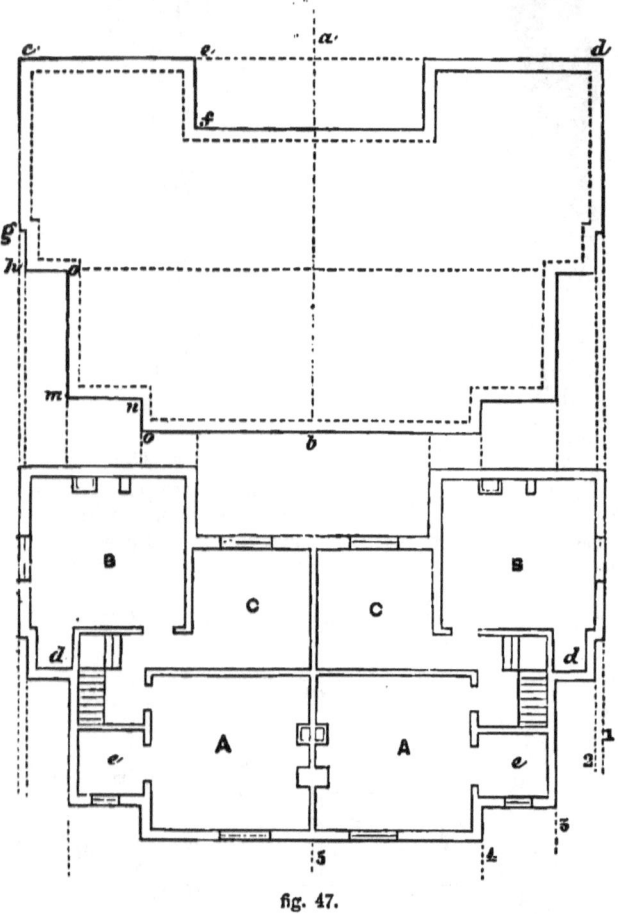

fig. 47.

fig. 46. Measure *o s*, fig. 46, and lay it off from *c* to *e*, fig. 47; at right angles to this draw *e f*, and make it equal to *s c* in fig. 46. Draw *c g* at right angles to *c e*, and make it equal to *o t* in fig. 46; make the short 'return' at *g* equal to that at *t* in fig. 46. Parallel to *g c* draw *g h*, and make it equal to *t u* in fig. 46; make the return *h o* at right angles to *g h*, and equal to that at *u* in fig. 46. At right angles to this draw *o m*, equal to *u v*:

make the return at *m n* equal to that at *v*, fig. 46; draw, parallel to *a b*, the line *n o*; make *o h* equal to *x d*, fig. 46. The other half, which is exactly similar, should be drawn in simultaneously with the first. After the outline is thus obtained, the thickness of the walls should next be put in, as shown by the dotted lines in fig 47. The example in fig. 47 is also designed to show the method of drawing a 'bedroom plan,' or floor above the ground one, from the data given by the lines on the latter. Suppose the upper figure (in 47) to be filled in with the partitions, fire-places, &c. &c., as in fig. 46, thus representing the ground-plan *finished*. By means of the T square produce all the boundary-lines of the upper figure to an indefinite distance on the paper below it, as shown by the lines 1, 2, 3, 4, 5; then proceed as before described in copying fig. 47 from the outline of fig. 46. The diagram will, it is hoped, be sufficiently explanatory of the method to be adopted, bearing in mind the lessons previously given. The pupil, in copying the various lessons given, should use a much larger scale than the limits of our pages will admit of. In the lower part of the figure 47, A is the principal bedroom, B the back bedroom, C the children's bedroom, *d* a small wardrobe, and *e* a small closet or bath room.

EXAMPLE 48. *To draw the plan of cellar in fig.* 48. Bisect *a b* in *c*, draw *c d;* corresponding to these, on the board draw lines *a b, c d*. From *c* measure to *a b*. Draw from these, at right angles to *a b*, to *e e;* parallel

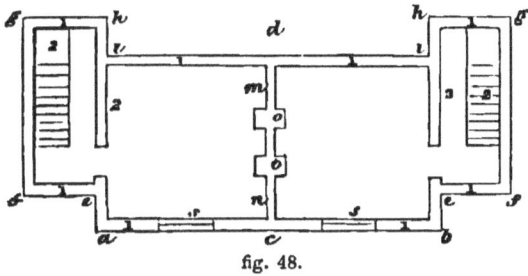

fig. 48.

to *a b* draw *e f*, and parallel to *c d, f g*. Parallel to *e f* draw *g h;* parallel to *c d, h i*. Join *i i;* the outline of the plan is thus obtained. Put in the thickness of the walls, the horizontal lines 1 1 first, the vertical 2 2 thereafter; and the central partition *m n*, with fire-jambs *o o*. Put in also the windows *s s*, and stairs, as in the drawing.

EXAMPLE 49 is designed to show the method of getting the position of the doors and windows in the front elevation, from the data afforded by the plan A P P E, fig. 49. The plan below represents the ground-plan of a row of four cottages, of which one-half is the counterpart of the other; we have, therefore, only shown the one-half fully drawn. The line G F, dividing the length into equal parts, is prolonged to H; the line *a b r* is drawn at right angles to this, and represents the ground-line: the distance of this above the plan will be decided according to circumstances, size of paper, &c. The openings of doors A, B, and E, are each bisected, and from the points lines are drawn parallel to G F, cutting the ground-line in the points *u, v*, and 3. In like manner, the windows C and D are bisected, and lines from the points drawn parallel to G F, cutting the ground-line in the points 1, 2.

The line 3 is the centre-line of end-door *p r*, the line 2 centre-line of window *n o*, line 1 centre-line of second window *h m*; the line *b*, of the window *d e*; *c*, of *f y*. The sizes of doors, &c., being previously ascertained, and

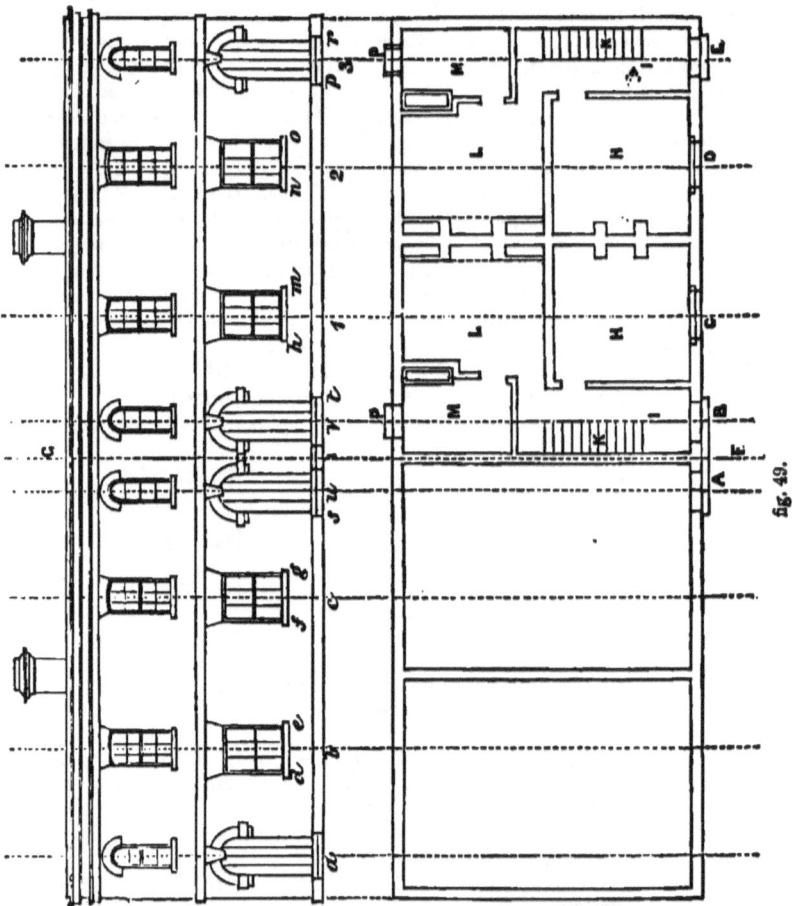

the scale known, the centre-lines obtained will enable the various parts to be drawn. In like manner, supposing the front elevation correctly drawn to scale given, also a rough sketch of ground-plan, with sizes, divide the length of front into two parts, and draw a line G at right angles to the ground-line. Draw any line parallel to the ground-line, at any distance below the elevation; this will form the back line of wall. Produce G to F; this will form the centre-line of the houses. Next bisect the breadth-line of doors in the points *a*, *u*, *v*, and 3; and from these points, parallel to G F, draw lines to A, B and E; next divide the windows *h m*, *n o* in the points 1, 2, and draw as before lines to C, D. From the points thus given, if the pupil

c

has carefully attended to the foregoing lessons, he will have no difficulty in drawing the various parts accurately. In the plan here given B and E are the principal doors, I I the lobby, K the stairs to bedrooms, H the living-room, L the kitchen, M the scullery, P the back entrance.

In the work on Practical Geometry we have amply illustrated the method of reducing irregular figures by means of squares; to that work, therefore, as introductory to the present, we refer the reader for information; we here content ourselves with giving, in

EXAMPLE 50, fig. 50, an architectural subject, having a series of squares

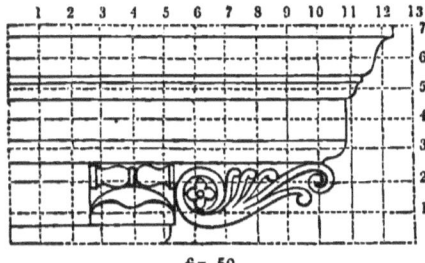

fig. 50.

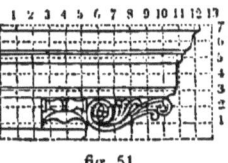

fig 51.

drawn over its surface, preparatory to its being reduced one-half, as shown in

EXAMPLE 51, fig. 51. Should it be required to enlarge fig. 50, all that is necessary is to draw the same number of squares, but of double the size, when, the various points being transferred to the proper places, an exact copy of fig. 50, but of twice the size, may be obtained.

In architectural drawing it is sometimes necessary to delineate the material of which the walls, &c., are constructed. Thus, in

EXAMPLE 52, fig. 52, a series of bricks built on one another is delineated. The bricks are so disposed as to 'break joint,' as it is termed;

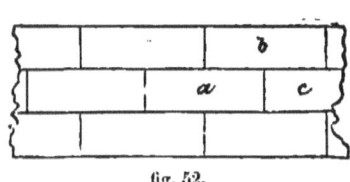

fig. 52.

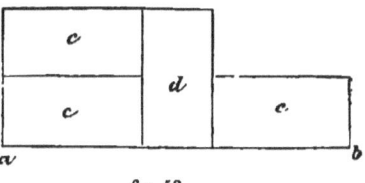

fig. 53.

that is, the solid part of b is placed over the joint formed by the juxtaposition of the bricks a and c. In ordinary work, bricks are used in two ways—as 'headers' and 'stretchers'—the 'headers' being placed across the wall, the 'stretchers' running along in the direction of its length. Thus, in

EXAMPLE 53, fig. 53, suppose $a b$ to be the line of wall, the bricks $c c c$ are 'stretchers,' and d a 'header.' The size of a brick of the ordinary dimensions is 9 inches long, $4\frac{1}{2}$ inches wide, and 3 inches thick. Brick work is generally laid in two kinds of bond, termed 'English' and 'Flemish' bond. By the term 'bond' is meant the tie between the various members of a brick wall, and which is generally secured by the proper disposition of the bricks; this is effected by the arrangement of the 'headers' and 'stretchers.' Thus, in

ENGINEERING, AND MECHANICAL DRAWING-BOOK.   35

EXAMPLE 54, fig. 54, which is a specimen of an elevation of a brick wall in 'English,' or as it is sometimes termed, 'old English bond,' where it con-

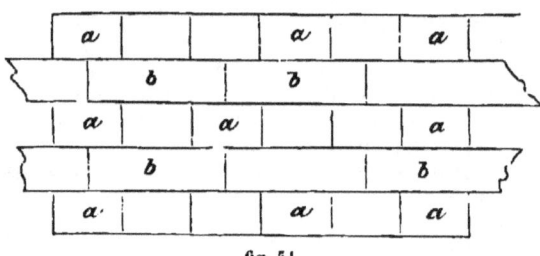
fig. 54.

sists of alternate layers of brick 'headers' and 'stretchers,' $a\,a$ being the 'headers,' and $b\,b$ the 'stretchers.'

EXAMPLE 55, fig. 55, shows a specimen of 'Flemish' bond, in which each row is made up of 'stretchers' and 'headers' laid alternately; $a\,a$ are

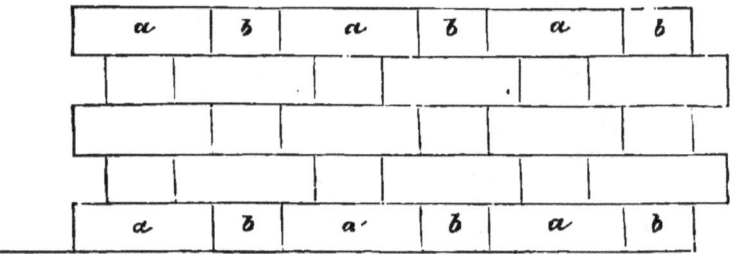
fig. 55.

the former, $b\,b$ the latter.   In delineating plans, various methods are in use for filling up.   Thus, in

EXAMPLE 56, fig. 56, a represents the method of filling up walls in a

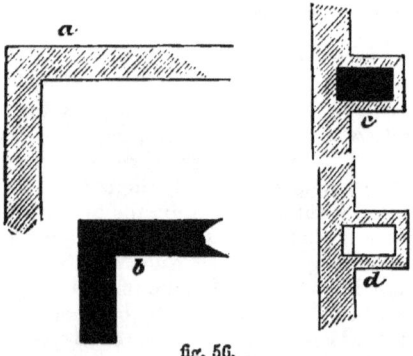
fig. 56.

plan by means of cross lines b where the whole is dark, all openings, at doors and windows, being left unshaded. The method of showing a chimney flue in the thickness of a wall is shown at c; another method in d. Stone work may be classed into three different kinds, as generally adopted; these are 'rubble,' 'coursed,' and 'ashlar.'

EXAMPLE 57, fig. 57, shows the method of delineating 'rubble work, in which the wall is composed of stones of all sizes and shapes.

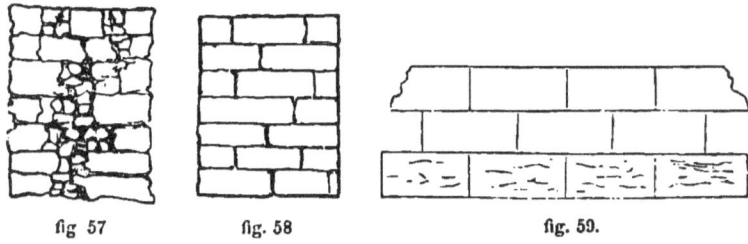

fig. 57   fig. 58   fig. 59.

EXAMPLE 58, fig. 58, shows the method of delineating 'coursed work,' in which the stones are, to a certain extent, squared and set in courses: hence the term.

EXAMPLE 59, fig. 59, shows the method of delineating 'ashlar work,' in which all the stones are squared up to certain given sizes, and set in regular courses.

EXAMPLE 60, fig. 60, shows the method of delineating 'vermiculated' work, in which the surface of the blocks is left with rough projections, a

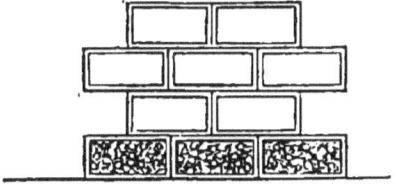

fig. 60.

narrow margin, tooled flat, being generally left round. This kind of work is used for 'keystones,' rusticated basements, doorways, &c.

The department now to be considered is that of

THE FIVE ORDERS—THEIR PROPORTIONS AND METHODS OF DELINEATION.

EXAMPLE 61, fig. 61, is an elevation of the 'Tuscan' order as generally received. The part from a to b is the 'pedestal,' from b to c the 'base,' from c to d the 'shaft,' from d to e the 'capital,' from e to f the 'entablature,'—the parts base, shaft, capital, and entablature, being termed a 'column.' The heights of the mouldings and the projections are all taken from the standard of measurement of each column; this standard being the diameter of shaft immediately above the base. This is divided into two equal parts, termed 'modules;' each of these again into thirty equal parts. The diameter is therefore divided into sixty equal parts; if necessary, each part is divided into sixty parts, called seconds. The standard is, therefore, thirty parts equal one module; two modules equal one diameter, or sixty parts. According to Palladio and other authorities, the height of column (Tuscan) now under consideration is, including base and capital, equal to seven diameters. To obtain, therefore, the diameter of any column, its height being given, all that is necessary is to divide the height into seven equal parts, one of which is the diameter; or where, on the contrary, the diameter is given, seven times this will give the height of column, including base and capital. We may now proceed to describe the laying out of the various members of a complete 'order,' showing the proportions of the mouldings, their height and projections. Although some writers discard the pedestal as an integral portion or a correct feature of any of the orders, we follow the majority of those who adopt it as a distinguishing feature. It is not here our province to enter into a detail of the æsthetic rules guiding the laying out of the various orders; we merely give examples of the parts as generally received. To those of our readers anxious to go into the matter, we refer to more technical works, or the treatise in this Series entitled *Ornamental and Architectural Design.*

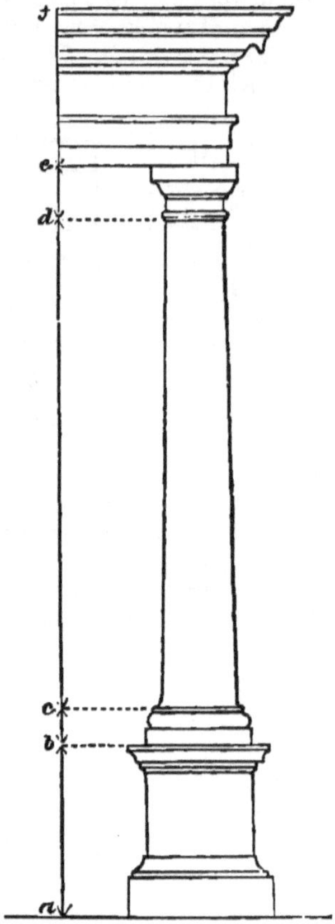

EXAMPLE 62. Suppose the line ab (fig. 62) to represent the diameter of a 'Tuscan' column. Dividing ab into two parts in the point c, ac, cb will be the two modules; dividing each module into three equal parts at $d, e, f,$ and g, and these again into five equal parts, a scale will be constructed from which to measure the various mouldings. Number as in the drawing.

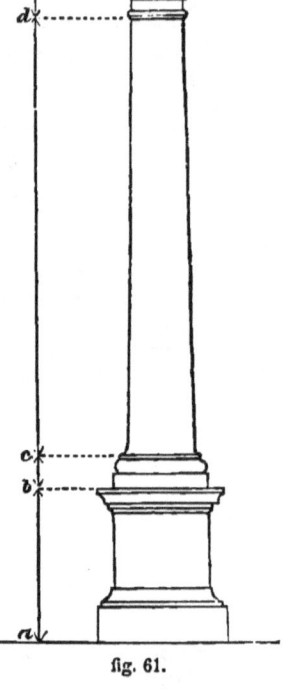

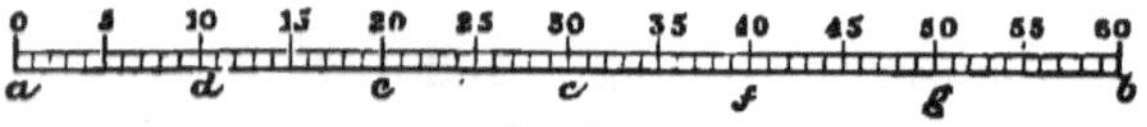

fig. 61.

fig. 62.

EXAMPLE 63, fig. 63, shows the method of proportioning the mouldings of the 'Tuscan pedestal.' Every pedestal is divided into three parts,—the 'base,' as A B; 'die,' B C; and the 'cornice,' C D. In the figure given the whole height of the pedestal is four modules, or ab, fig. 62. In order to keep our sketches within the limits of the page, we take the proportions from a scale, the divisions of which are only half the size of those in fig. 62. At a, fig. 63, draw a line, as iai, of indefinite length, and at right angles to it a line ab; make ab equal to 2 diameters, or 4 modules, ac equal 26 parts, cd equal 4 parts, de equal 8. Make the 'die' en equal 2 modules 4 parts; make bf equal 3 parts, fg equal 8, gh equal 2, gn equal 4: the heights of the 'members' will thus be found. The projections of the mouldings are all set out from the central line ab. From a with 53 parts lay off to i, i, and from these draw lines meeting that drawn from c; parallel to iai make

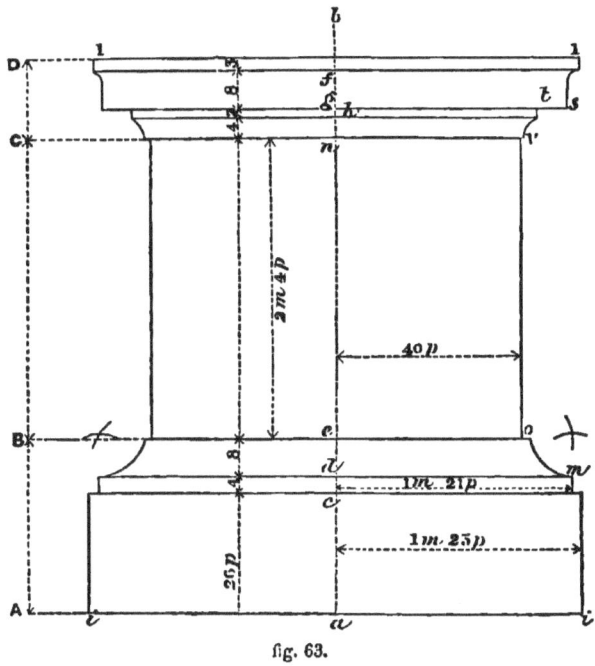

fig. 63.

dm equal 51 parts, or set back the line md 2 parts from the end of line c: make eo equal $41\frac{1}{2}$, and the die equal 40 parts; make b 1 equal 53 parts; make gs equal $50\frac{1}{2}$, and st equal 7, and nv equal eo.

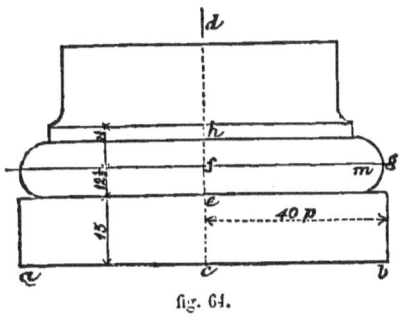

fig. 64.

EXAMPLE 64, fig. 64, shows the 'base' of the Tuscan order. Draw the centre-line cd, put the 'plinth' ab, making cb equal 40 parts, and ce equal 15; make the 'torus' moulding in height equal $12\frac{1}{2}$ parts. The centre m of the circular termination is found on the line f. Make the fillet h equal $2\frac{1}{2}$ parts, and its projection from centre-line equal $33\frac{3}{4}$, or nearly 34 parts. To describe the 'apophygee,' by which the lines of shaft are connected with the base, see work on Practical Geometry, where also the various forms of mouldings met with in the Orders may be found described, and the methods of delineating them.

EXAMPLE 65, fig. 65, is the Tuscan 'capital,' drawn to the same scale as the others. Draw cd, ab at right angles; make ca, cb equal $22\frac{1}{2}$ parts, or ab equal 45; make the fillet of the astragal en equal $24\frac{1}{2}$ parts, or nen equal 49 parts. Make gh equal 27; gi, the 'neck,' equal ab or 45 parts, and the fillet m above the neck equal en. Make the diameter of 'abacus' $n'o$ equal 60 parts, or 1 diameter. These are the projections; the heights are as follows:—The fillet ef equal 2 parts; fg equal 4; gm equal $8\frac{1}{2}$; the fillet above this $1\frac{1}{2}$; the quarter-round mn' equal 10; the abacus or plinth $n'o$ equal 10. The quarter-round begins at 1 division of the scale from s.

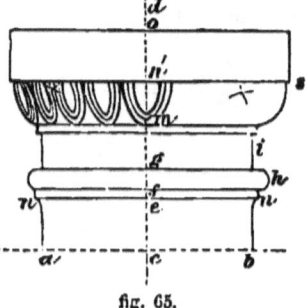

fig. 65.

EXAMPLE 66, fig. 66, is an elevation of the Tuscan 'entablature.' Every entablature consists of three parts,—the 'architrave' A B, the 'frieze' B C, the 'cornice' C D. Draw the line bd representing the centre line of column,

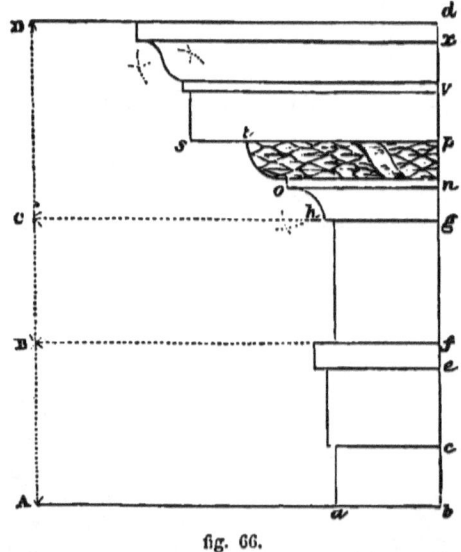

fig. 66.

and ab at right angles to it. The connection of the entablature with reference to the column will be seen in fig. 61. In the present figure the position is reversed. Make bc, the lowest 'fascia,' equal $12\frac{1}{2}$ parts in height and $22\frac{1}{2}$ in width from the central line bd to a. The upper 'fascia' cc is 17 parts in height and 24 in width; the 'fillet' ef is 5 parts in height and $27\frac{1}{2}$ in projection; the height of the 'frieze' fg is 26 parts, and its projection $22\frac{1}{2}$; the first moulding in the cornice gn (the cavetto), equal $7\frac{1}{2}$ in height, and projection gh equal 24. Make the fillet equal $1\frac{1}{2}$, and its projection no equal 32; make the height of 'quarter-round' from n to p equal 9, and its projection ps equal $52\frac{1}{4}$: make pt equal 40, and join ot; make the 'corona' pv equal 10 in height, and the 'fillet' above it equal 2; its projection equal $54\frac{1}{4}$. Put in the 'cyma recta' $to\ x$, equal 10 parts, the last fillet equal $3\frac{1}{2}$, and its projection equal 66.

EXAMPLE 67, fig. 67, shows the elevation of the 'Doric column,' with 'pedestal' ab; bc the 'base,' cd the 'shaft,' df the 'capital,' and fe the

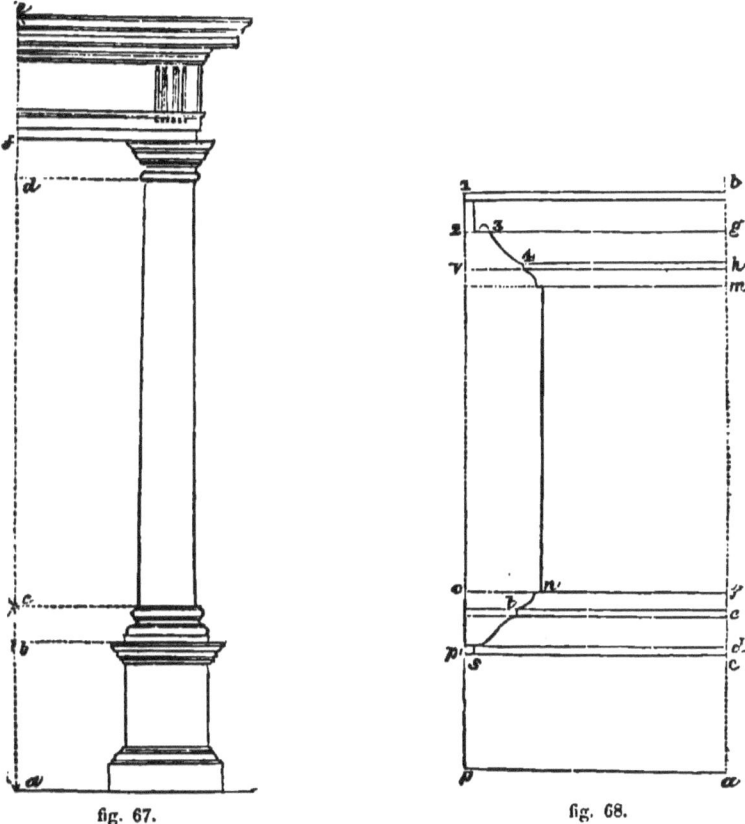

fig. 67.   fig. 68.

'entablature.' The height of the column, including base and capital, is equal to seven diameters.

EXAMPLE 68, fig. 68, is the elevation of half of the pedestal of the Doric column to same scale as the last example. Draw $ap\ ab$ at right angles, make ab equal to 4 modules 5 minutes, or 4 modules 20 minutes. Make the 'plinth' ac equal 26 parts in height; the 'fillet' cd equal $1\frac{1}{2}$; the 'cyma recta' de equal $6\frac{1}{2}$; the 'fillet' e equal 1; the 'cavetto' f equal 4. Proceed now to put in the cornice; make the top 'fillet' at b equal 2 parts; the 'corona' below equal $6\frac{1}{2}$; the 'quarter-round' equal $6\frac{1}{2}$; the 'fillet' equal 1, and the 'cavetto' equal 4. Put in the breadth of the 'die' by measuring from f to n, equal 40 parts. From n, the face of the die, measure off to o, equal 16 parts; through o draw a line to p parallel to $a\,b$. From p' set off to s, equal 2 parts; from the line $p\,o$ to t, equal 11 parts. Make the projection of the cavetto at top of base and at cornice equal to 1 part from line of die. From v lay back to 4, equal 12 parts; from 2 to 3, equal $5\frac{1}{3}$. Put in the cyma at $s\,t$, and the quarter-round from 4 to 3.

EXAMPLE 69, fig. 69, represents the base of the column now under consideration; it is sometimes termed the 'Attic base;' 10 parts are given to the 'plinth;' 7 to the 'torus;' $1\frac{1}{2}$ to the 'fillet;' 4 to the 'scotia;' 1 to the

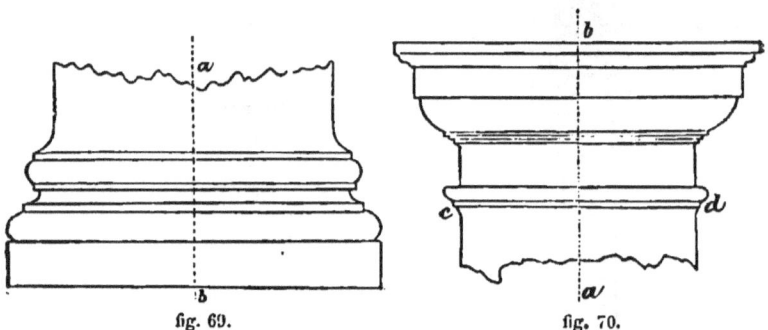

fig. 69.   fig. 70.

fillet above it; $5\frac{1}{3}$ to the second torus, and 1 to the fillet above. The projections are set off from the centre-line $a\,b$, and are as follows, commencing with the 'plinth' equal 40; 'torus' equal 40; 'fillet' equal $36\frac{2}{3}$; fillet beneath the second torus 35; second torus $36\frac{2}{3}$; last fillet 34.

EXAMPLE 70, fig. 70, is the 'capital' of the Doric order. The various 'heights' and 'projections' are as follows, beginning with the fillet $c\,d$. The diameter of top of shaft is 52, or 26 parts on each side of the centre-line $a\,b$; fillet $c\,d$ is $1\frac{1}{2}$ parts in height, and projection 28; the astragal or bead $3\frac{1}{2}$, projection 30; the neck 9 parts, projection 26. The three fillets below the quarter-round are together $3\frac{1}{3}$ parts in height; this is divided into three equal parts, as in the drawing. The quarter-round is $6\frac{1}{2}$ in height; the 'abacus' $6\frac{3}{4}$, and its projection 36: the quarter-round below it begins at a point 1 part back from end of abacus; the last fillet is 39.

EXAMPLE 71, fig. 71, shows an elevation of the Doric entablature. The line $x\,x$ is the centre-line of column (see fig. 67), from which the projections are taken. The architrave $a\,f$ is composed of two fasciæ $a\,b$, $b\,d$, with a fillet $d\,f$. The 'guttæ' or 'drops' in the upper fascia $b\,d$ are $3\frac{2}{3}$ parts in height, surmounted with a fillet $1\frac{1}{4}$. The 'triglyph' is over this in centre of column, and its width is 30 parts; the distance between the 'triglyphs' is exactly a square, the side of which is the depth of the frieze $f\,g$; the distances between the triglyphs are called 'metopes,' and are generally filled in with some ornament as in the drawing. The following are the heights of the various mouldings, with their projections: $a\,b$ 11 parts, projection from $x\,x$ equal 26; $b\,c$ equal $9\frac{1}{2}$; $c\,d$ equal $3\frac{2}{3}$; $d\,e$ $1\frac{1}{3}$, the projection of b and d equal 27; of the fillet $f\,d$ equal 28, its height being 4 parts; the height of frieze $f\,g$ equal 45; $g\,h$ equal 5, projection of $g\,h$ equal 27. Height of $h\,k$

equal 5, the fillet 1; projection of h equal 32; of k 35½; height of $k\,m$ equal 6; projection of m equal 64½; of $v\,t$ equal 39½. Height of $m\,n$ equal 8;

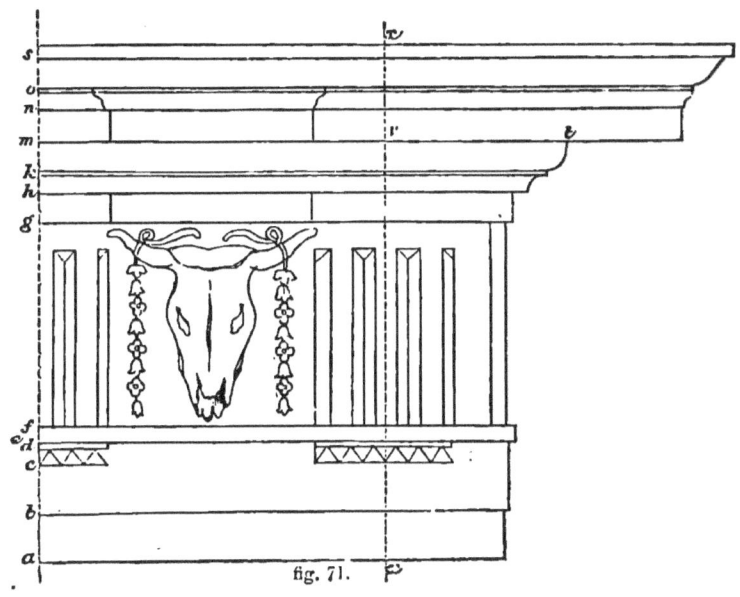

fig. 71.

fig. 72.

$n\,o$ 3¼; the fillet ¾, its projection 68. Height of $o\,s$ equal 6¾; fillet equal 2¼; projection 76. The method of drawing the 'triglyphs' and 'guttæ' of this order is further elucidated by

EXAMPLE 72. Let $a\,b$ (fig. 72) be the height of 'frieze,' and $c\,d$ semi-diameter of column at base. Make $b\,e$ equal 4 parts; the fillet $e\,e'$ beneath this equal 2; and from e' to f equal 4. Divide $c\,b$, $b\,d$ each into six equal parts; and parallel to $a\,b$, draw through these lines as in the drawing to the line $g\,h$. On $g\,c$, $h\,d$, lay off equal 2½ parts to m, m; and with $m\,n$ from m, lay off to o; join $n\,o$, $n\,o$. On the lines 4 4 draw to $o\,o$, and put in the angular lines. Bisect the fillet $b\,e$ in the line $s\,s$; from the points 1, 2, 3, &c. at f, draw lines to $s\,s$ where this line intersects the vertical ones, dotted as in the sketch. These angular lines are only continued to the under side of fillet e'.

EXAMPLE 73 represents the elevation of the 'Ionic' order. A, fig. 73, is the base of pedestal, B the die, C the cornice, D the base of column, E the shaft, F the capital, G the architrave, H the frieze, I the cornice of entablature.

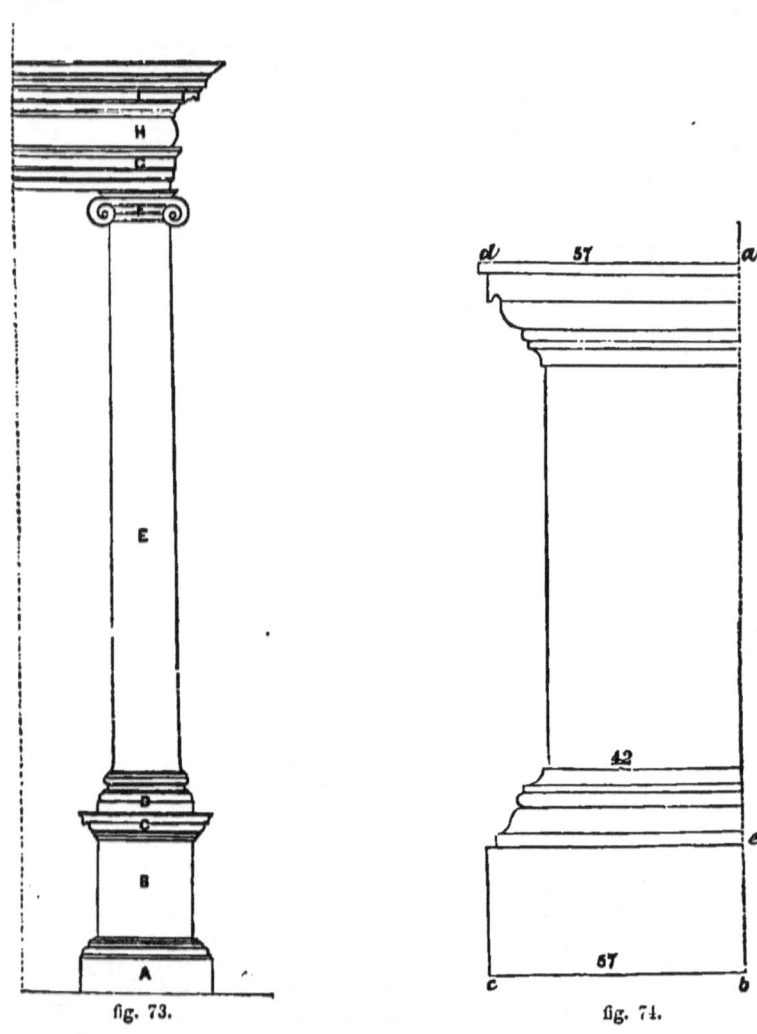

fig. 73.   fig. 74.

EXAMPLE 74, fig. 74, shows the elevation of half of Ionic pedestal; the line ab being that from which the projections are taken; the plinth bc is $28\frac{1}{2}$ parts in height, and 57 in projection. The upper fillet ad is 2 parts high, and 57 in projection. The width of die is 42 parts. The whole height of pedestal from a to b is two diameters 34 parts, or 4 modules 4 parts. The heights of the other mouldings and projections are as follows, commencing with the fillet at *e* above the plinth, which is in height 1½ parts, projection 54½; the cyma 6½ in height, projection 48½; the astragal 2½ in height, projection 50; the fillet 1, projection 48½; the cavetto, 3½, projection 43. The height of die 87 parts; the height of cavetto above die 4 parts, projection 43; the fillet 1, projection 46; the astragal 3½; projection 48; the quarter-round 6; the corona 6, projection 55.

EXAMPLE 75, fig. 75, is the Ionic base, the line *a b* being the centre-line. The heights and projections are as follows: the plinth *c d* 10 in height, 42 projection; the torus, 8 height, 42 projection; fillet, 1 height, 37 projection; scotia, height 5; second fillet, height 1, projection 34½; second torus, height 5, projection 37; astragal, height 2, projection 34½; thi fillet, height 1½; projection 33.

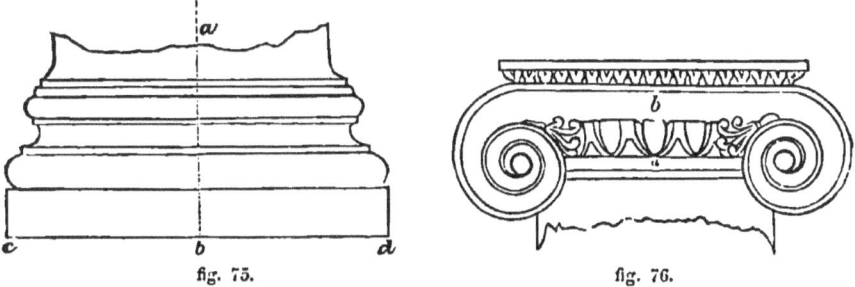

fig. 75.   fig. 76.

EXAMPLE 76, fig. 76, shows the elevation of Ionic capital drawn to same scale as the others. The plan of the capital is shown in

EXAMPLE 77, fig. 77, and the side view in

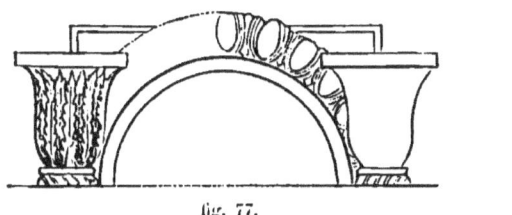 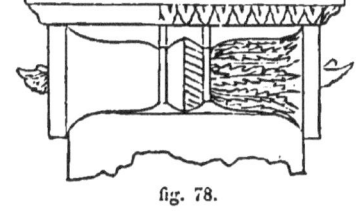

fig. 77.   fig. 78.

EXAMPLE 78, fig. 78. The method of describing the scroll termed the 'volute' is explained in

EXAMPLE 79, fig. 79. Draw *a b*, *c d* at right angles; let *e f* be the diameter of the eye of the volute corresponding to the breadth of the astragal *a* (see fig. 76); with half *ef* from the point where *a b*, *c d* intersect, describe a circle; within this inscribe a square. In fig. 80 the centre of the volute is drawn to a larger scale, to enable the pupil to mark out the centres used to describe the scroll in fig. 79. From *e*, fig. 80, with radius *e d*, describe the circle, and within it inscribe the square *a b d c* corresponding to the square *e g f h* in fig. 79. Through *e*, the centre, parallel to *c a* draw *f h*, and parallel to *a b*, *i g*; join the extremities, and form a square *i h g f*. Divide the diagonals *i g*, *f h* each into six equal parts, at the points 1, 2, 3, 4, 5, 6, 7, 8. At

ENGINEERING, AND MECHANICAL DRAWING-BOOK. 45

these points draw lines at right angles, forming squares of which the corners are only given in the diagram to avoid confusion. Divide $i k$ into four equal parts; from h lay one of these to m; from i to n; from f to o; from g to p; from 8 to s; from 1 to t; from 5 to v; from 4 to x; from 7 to y; and so

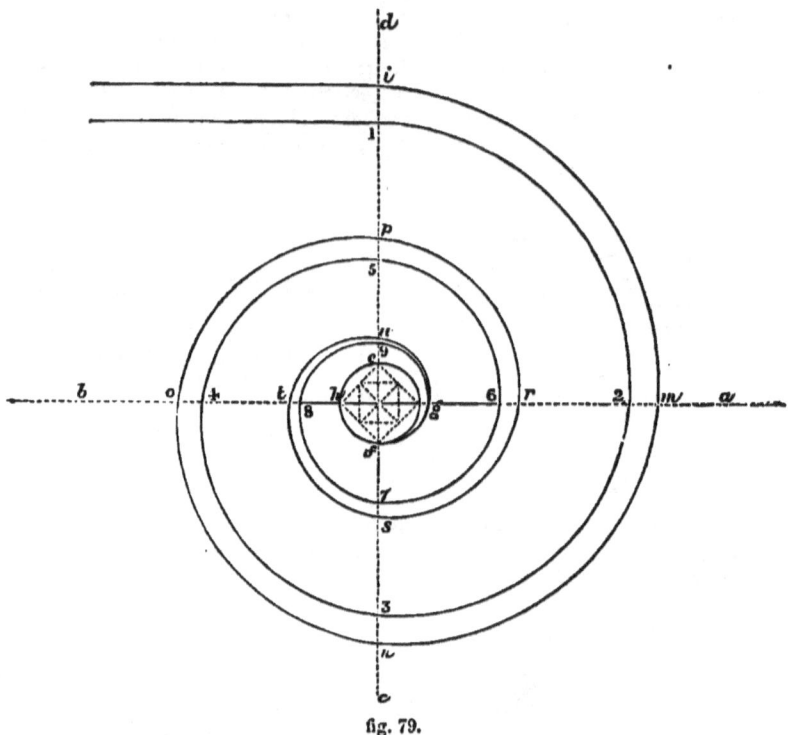

fig. 79.

on to the point of the square corner at 3. These various points thus obtained are the centres from which the curve is described. Suppose the point i, fig. 79, to be the under line of abacus of capital, as b (see fig. 76), from the centre, on line eh, fig. 79, corresponding to the point c, fig. 80, with radius hi describe an arc of a circle to the point m, meeting the diameter of gh prolonged to a. From the point in the smallest square in fig. 79, corresponding to the point a, fig. 80, with radius em, fig. 79, describe an arc mn, meeting the diameter ef prolonged to c. From the point on the small square, fig. 79, corresponding to b, fig. 80, as a centre, with gn as radius, describe an arc no, meeting gh produced to b. From f as centre, with fo describe an arc to p, meeting line cd. From centre 1 (see fig. 80), with radius $1p$ describe an arc to r. From centre 8 (see fig. 80), with 8 r as radius, draw an arc to s. From centre 4 (see fig. 80), with 4 s describe an arc to t; from centre 5, with radius 5 t, describe an arc to w; from centre 2 (see fig. 80), with radius 2 w describe an arc to g, and so on. To draw the interior curve proceed as follows: from the point n (see line if, fig. 80), with radius $m1$, describe an arc to the point 2 in the line *a b*, fig. 79; from the point *m* (see line *i h*, fig. 80) with the radius *m* 2, an arc to the point 3 on the line *c d*, fig. 79; from the point *p* (line

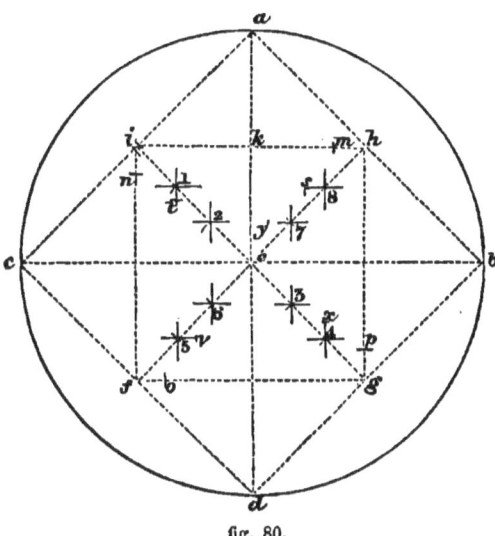

fig. 80.

h g, fig. 80), with the radius *p* 3, an arc to 4; from the point *o* (line *g f*, fig. 80), with radius *o* 4, an arc 5, and so on from the centres corresponding to the points *s*, *t*, *v*, *x*, *y*, &c., describing curves to the points 5, 6, 7, 8, 9, &c. fig. 79.

EXAMPLE 80, fig. 81, represents the 'Ionic entablature;' *a b* being the centre-line of column, and that from which the projections of the various members are taken. In succession, beginning from the point *b* upwards, the heights and projections of the various mouldings are as follows:—

1st height equal		$6\frac{1}{2}$ parts,		projection equal		$26\frac{1}{2}$
2nd ,,	,,	2	,,	,,	,,	27
3rd ,,	,,	8	,,	,,	,,	$27\frac{1}{2}$
4th ,,	,,	$2\frac{1}{2}$	,,	,,	,,	29
5th ,,	,,	10	,,	,,	,,	$29\frac{1}{2}$
6th ,,	,,	5	,,	,,	,,	33
7th ,,	,,	3	,,	,,	,,	35
8th ,,	,,	27	,,	,,	,,	33
9th ,,	,,	5	,,	,,	,,	27
10th ,,	,,	1	,,	,,	,,	32
11th ,,	,,	6	,,	,,	,,	36
12th ,,	,,	2	,,	,,	,,	37
13th ,,	,,	7	,,	proj. to *e* equal 38,		to *c* equal 52
14th ,,	,,	3	,,	,,	,,	55
15th ,,	,,	4	,,	,,	,,	60
16th ,,	,,	4	,,	,,	,,	$63\frac{1}{2}$
17th ,,	,,	1	,,	,,	,,	64
18th ,,	,,	7	,,	,,	,,	64
19th ,,	,,	2	,,	,,	,,	72

EXAMPLE 81, fig. 82, represents an outline sketch of the 'Corinthian column,' with pedestal complete. The height of column is $9\frac{1}{2}$ diameters, including base and capital. A is the base of pedestal, B the die, C the cornice, D the base of column, E the shaft, F the capital, G the architrave, H the frieze, I the cornice.

EXAMPLE 82, fig. 83, is the pedestal of the Corinthian order. The proportions are as follows, taking them in their order from $b\,c$: the plinth $b\,c$, $23\frac{1}{2}$ parts in height, its projection from the central line $b\,d$ to a 57 parts;

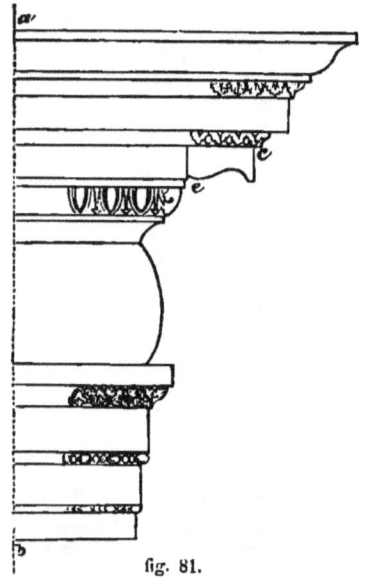

fig. 81.

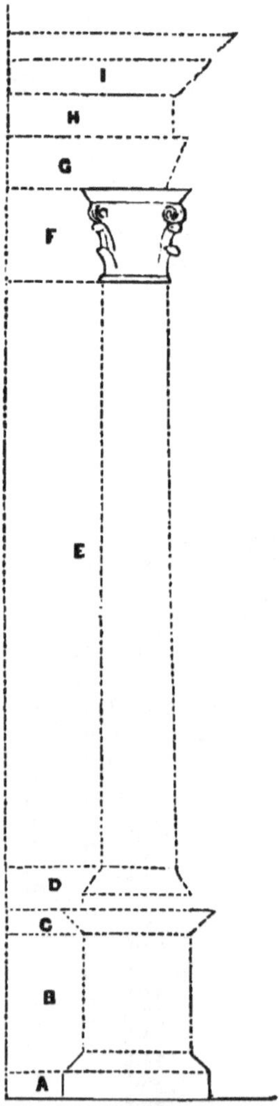

the torus, height 4, projection 56; fillet $\frac{3}{4}$, projection 55; cyma 5, projection 47; fillet 1, projection 47; cyma $3\frac{1}{2}$, projection 42; die 3 modules $4\frac{1}{2}$ parts; projection of die, 42 parts; the cavetto in cornice $3\frac{3}{4}$, projection 43; fillet $\frac{3}{4}$, projection 46; quarter-round $4\frac{3}{4}$, projection 50; corona $4\frac{1}{2}$, projection 53;-cyma $3\frac{1}{2}$, projection 57; the top-fillet $2\frac{1}{2}$, projection 57.

EXAMPLE 83, fig. 84, represents the base of the order, of which ab is the centre line. The heights in the progression of their order, commencing with $b\,c$, are as follows: 10, 7, 2, 1, 4, $\frac{1}{2}$, 2, 6, $2\frac{1}{2}$, 2. The projections, beginning with $b\,d$, are as follows: 42, 42, 38, 37, 32, 37, 35, 32.

EXAMPLE 84, fig. 85, represents the capital of the order. The diameter of shaft at the neck is $52\frac{1}{2}$ parts; the fillet $1\frac{1}{2}$, its projection 56; the astragal 4, projection 60. The height from a to b is 70 parts, the projection from b to c 46, the projection from b to e 60. Join ef, prolong af, be to g and h: join gh by a line parallel to ba, and mark off on it from gh, as in the sketch. From the points obtained draw

fig. 82.

lines parallel to be; the intersection of these with ef will give the position of the acanthus leaves. The method of laying out the plan of this capital is shown in fig. 86, where ab is the diameter of shaft at neck, ce corresponding to the distance bc, fig. 85. The centre of the circle of which dd is a part, is found by the intersection of the lines at f.

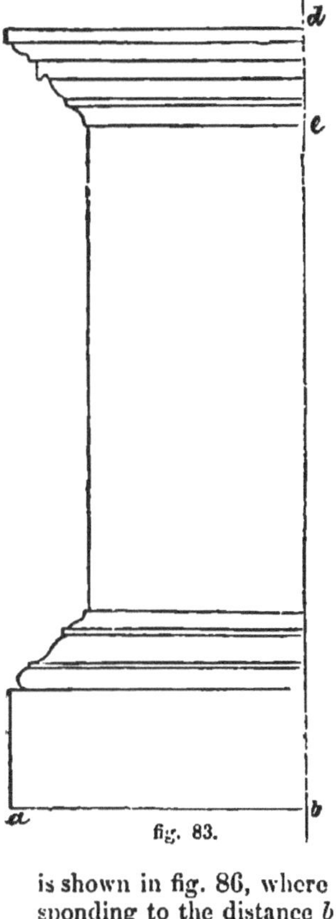

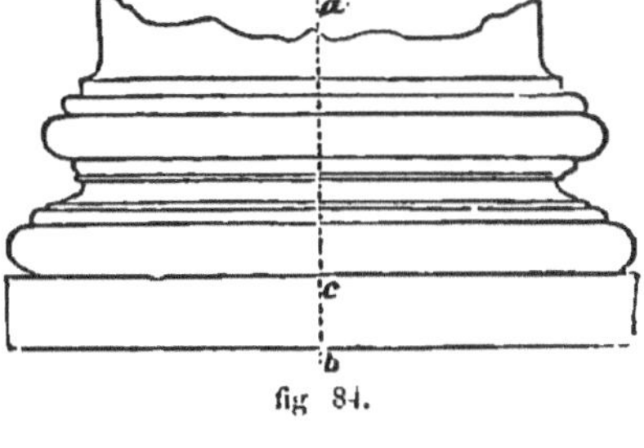

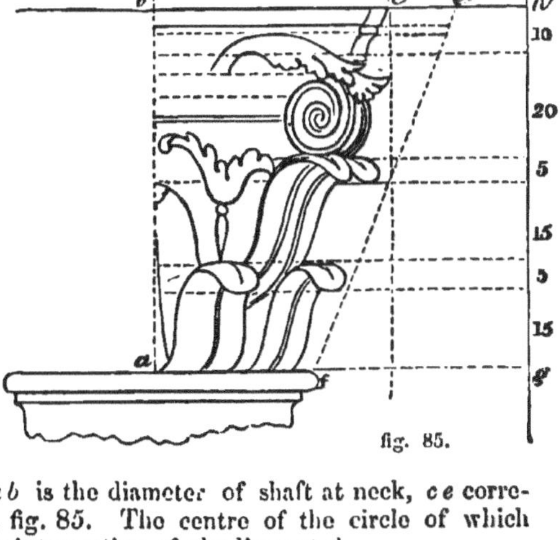

EXAMPLE 85, fig. 87, shows a form of capital of this order, with the ornaments filled in.

EXAMPLE 86, fig. 88, is the 'Corinthian entablature.' The heights of the different mouldings, commencing with ab, are as follows: 6, 1¾, 8¼, 1¾, 10½, 5, 2½, 28½, 4½, 1, 5½, 1, 4½, 1, 7½, 2½, 1, 7½, ⅔, 3, ¾, 6, 2¼. The projections, beginning with ac, are as follows: 26, 26½, 27, 27½, 28, 29½, 34½, 26, 26½, 32, 34, 35, 40, 58½, 60, 62, 62½, 66, 74.

EXAMPLE 87, fig. 89, represents the outline of the Composite order with pedestal complete: the letters and parts correspond

fig. 87.

with those given in fig. 82, where the pedestal is delineated. Its height, including base and capital, is 10 diameters.

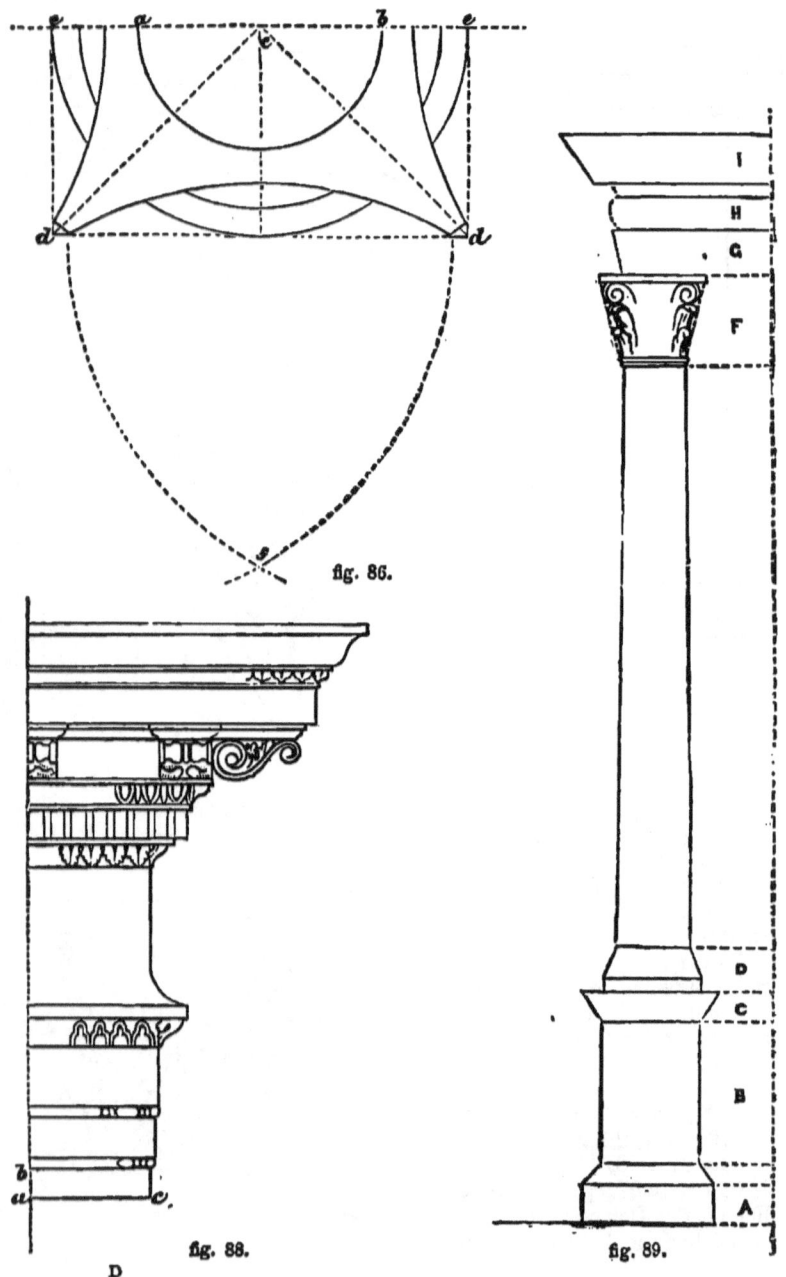

fig. 86.

fig. 88.

fig. 89.

EXAMPLE 88, fig. 90. The pedestal. The heights, commencing with bc, are as follows: 33, $4\frac{1}{2}$, 1, 3, $1\frac{1}{2}$; 1, height of die de 4 modules 5 parts. The height of mouldings in cornice, beginning at e, are as follows: $1\frac{1}{2}$, 3, $8\frac{1}{2}$, 1, $5\frac{1}{2}$, $3\frac{1}{2}$, $2\frac{1}{2}$. The projections, beginning with ab, are 57, 57, 55, 46, 45, 42, $44\frac{1}{2}$, 47, $52\frac{1}{2}$, 53; top-fillet 57.

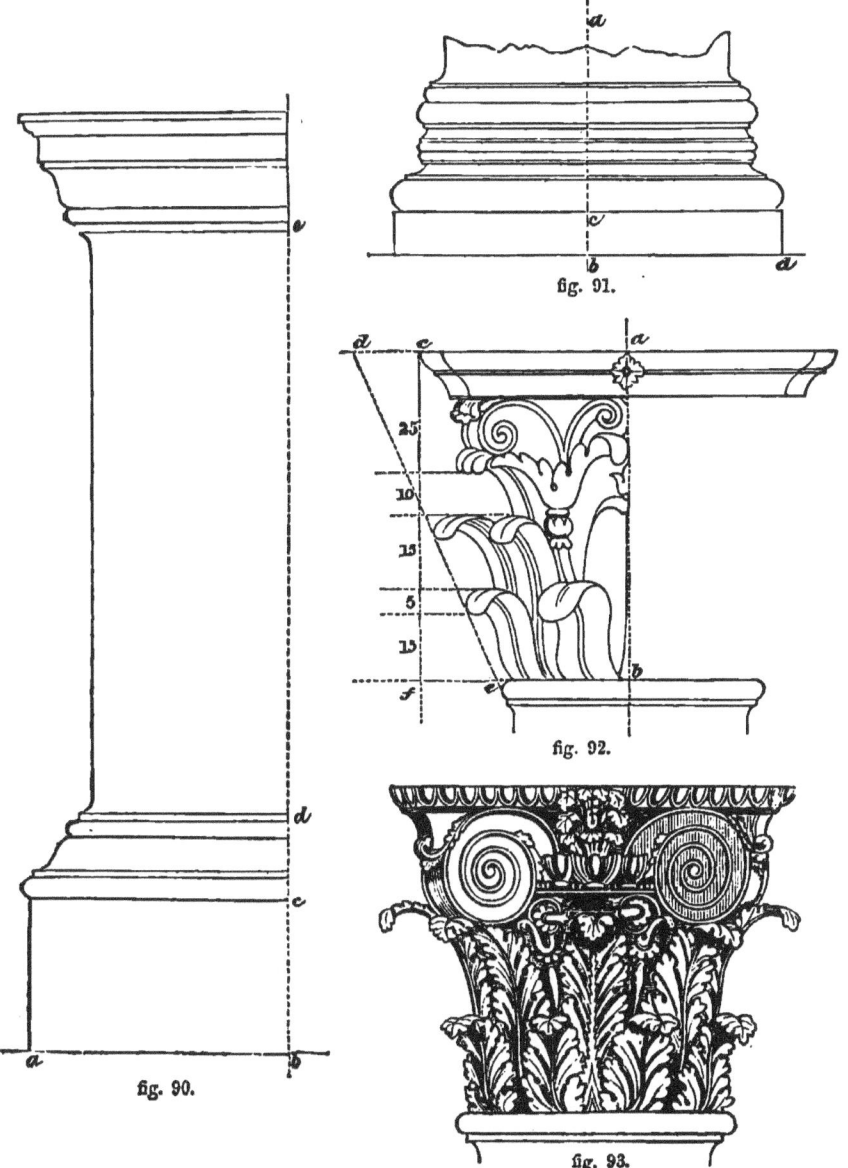

fig. 91.

fig. 92.

fig. 90.

fig. 93.

EXAMPLE 89, fig. 91, represents the base of the order. Heights, beginning with bc, 10, 7, $\frac{1}{2}$, $2\frac{1}{2}$, $\frac{1}{2}$, 2, 2, $\frac{1}{2}$, 2, $\frac{1}{2}$, $4\frac{1}{2}$, 2, 1; projections, beginning with bd, 42, 42, 38, 36, 37, 37, 36, 36, 37, 36, 34.

EXAMPLE 90, fig. 92, represents the capital of the order. The semidiameter of shaft at neck is 26 parts; the fillet $\frac{1}{2}$ in height and 27 in projection; the astragal 4 in height and 29 in projection. The height from b to a is 70, projection from a to c 45, and to d 60; the heights on the line fc are used by the intersection of the line dc to find the height of the ornament. (See example 84.) Another form, with the ornaments filled up complete, is given in

EXAMPLE 91, fig. 93.

EXAMPLE 92, fig. 94, is the Composite entablature. The heights of the mouldings, beginning with bc, are as follows: 12, $2\frac{1}{2}$, 15, $1\frac{1}{2}$, $3\frac{1}{2}$, 4, 2, 30,

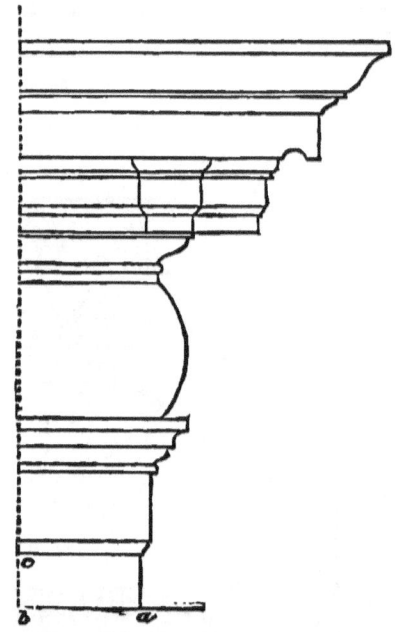

fig. 94.

2, 2, 5, $\frac{1}{2}$, $3\frac{3}{4}$, $1\frac{3}{4}$, $6\frac{1}{2}$, $1\frac{1}{4}$, $2\frac{1}{2}$, $9\frac{1}{2}$, $3\frac{3}{4}$, 1, 8, $2\frac{1}{4}$. The projections, beginning with ba, are 26, 28, 29, 34, 37, 36, 32, 52, $53\frac{1}{2}$, 54, 55, 66, 67, 70, 78.

The next example shows the manner of delineating intercolumniations. By this term is meant the distance between two columns, as a and b.

EXAMPLE 93, fig. 95, which is the intercolumniation of the Tuscan order. The distance between the columns is 6 diameters, the general distance, however, being 4 diameters. The pupil, at this stage of his proceedings, should make drawings to a large scale, as of that in fig. 62 of the intercolumniation of all the orders, to assist him in which we here give the various distances for each. The distance between the Doric columns is equal to three diameters; the distance in the Ionic is two diameters and a quarter; the distance between the columns in the Corinthian is two diameters and a quarter; and that of the Composite one diameter and a

fig. 95.

quarter to one diameter and three quarters. For the various species of intercolumniation, with their distinguishing names, see the work in this Series on *Ornamental and Architectural Design*.

Where it is necessary to introduce doors, windows, &c., thus widening the space between the columns to a greater extent than true proportion requires, 'coupled columns' are introduced, the distance between them being such as to allow of the proper projection of their 'capitals.'

EXAMPLE 94, fig. 96, shows coupled columns in the Corinthian order, where the space between the two columns is a little over two diameters.

Pilasters bear a considerable resemblance in their elevation to columns. The height of members and their projections are the same as the columns of the same order; the plan, however, instead of being circular as in columns, is square, the external surface being flat.

fig. 96.

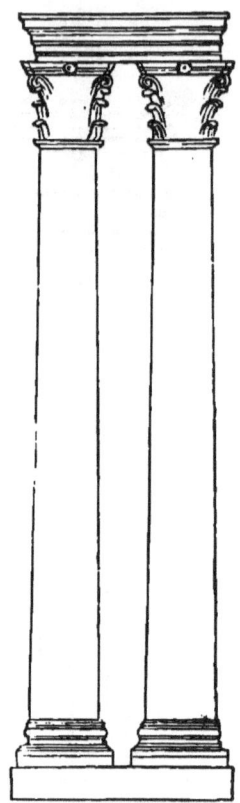

fig. 97.

EXAMPLE 95, fig. 97, shows 'coupled pilasters' in the Corinthian order.

Caryatides are sometimes used in place of columns and pilasters. These are representations of the human figure. When female, they are known by the name as above; when male, as 'Persians.'

EXAMPLE 96, fig. 98, is an exemplification of a caryatides. As a series of columns at proper distances form a colonnade, so columns with arches between them, are termed *arcades*. The Tuscan arcade is given in

EXAMPLE 97, fig. 99. The distance between the columns a and b is six diameters; A is termed a 'pier,' B the 'impost,' C the 'archivolt,' and D the 'keystone.' A semi-diameter of column is laid from c to d, which gives the line of pier $h\,d$. The distance from p to t is six diameters and three-quarters; a line through t parallel to $a\,b$ gives the height of im-

fig. 98.

post; the capital of impost is obtained by dividing gh into seven or eight equal parts, and giving one of these from m to n; the width of archivolt so is equal to one-ninth part of gh; the width of keystone at ef is equal to os. By drawing lines to e and f from t, the centre of the circle, $mefs$, the diverging lines will be obtained. To assist the pupil in making out

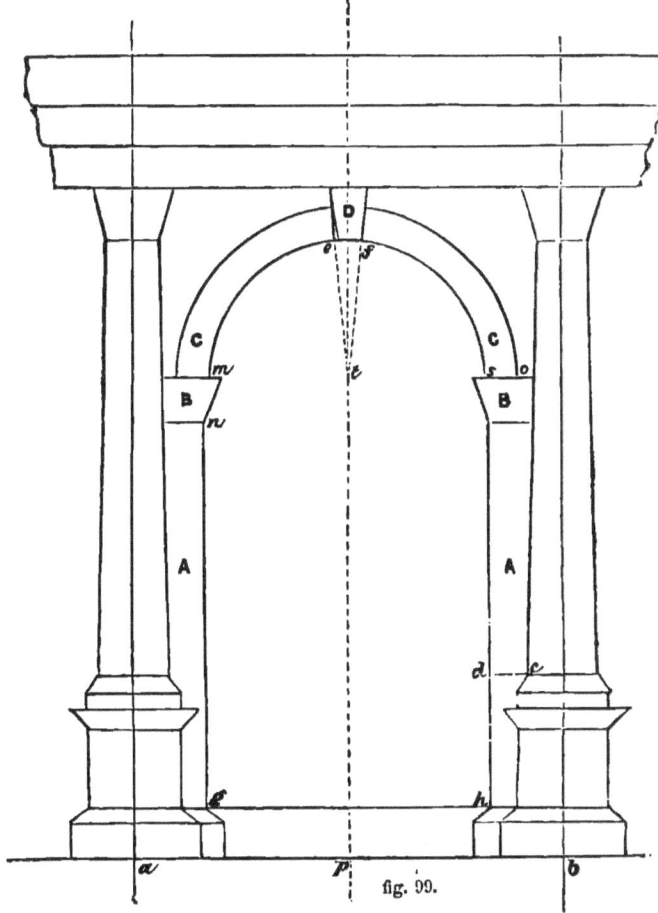

fig. 99.

examples of arcades in the other orders, we quote the following directions of a celebrated author on architecture as to proportions:—"The height of arches to the underside of their crowns should not exceed twice their clear width, nor should it be much less; the piers ought not to be less than one-third the breadth of the arch, nor more than two-thirds." The pupil desirous of studying the principles of architectural design may consult the work on *Ornamental and Architectural Design*, above noticed.

EXAMPLE 98, fig. 100, is an elevation of the Tuscan impost, with the heights and projections. The projections are set forward from b to k, in the line bc, the line bc representing the face of pier corresponding to the line hd in fig. 99. The scale from which the measurements are taken is that in fig. 62. The figures 1, 2, and 3 denote the width of the mouldings on the archivolt c c (see fig. 99), and are set back on the line ak from b.

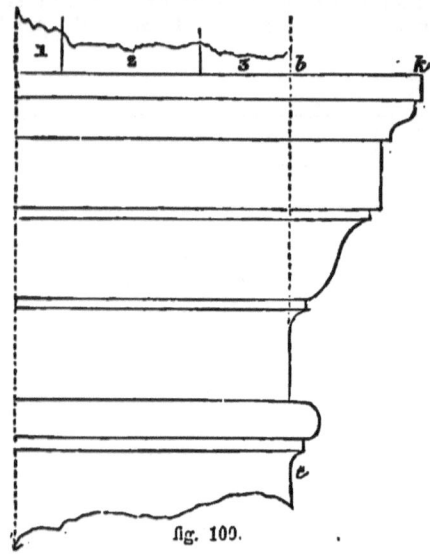

fig. 100.

EXAMPLE 99, fig. 101, is the Doric impost. The heights are measured from the point b on the line bc representing the line of pier, as in last ex-

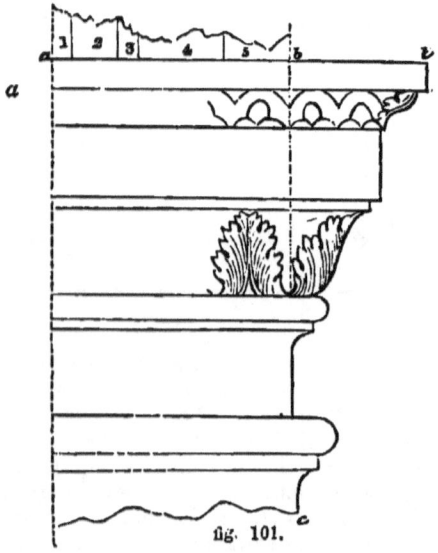

fig. 101.

ample, the projections being set forward from b to w and t, the width of mouldings of archivolt, 1, 2, 3, 4, 5, being from b towards a.

EXAMPLE 100, fig. 102, is the Ionic impost, the projections, heights, and widths 1, 2, &c. of archivolt mouldings being set out as in last figure.

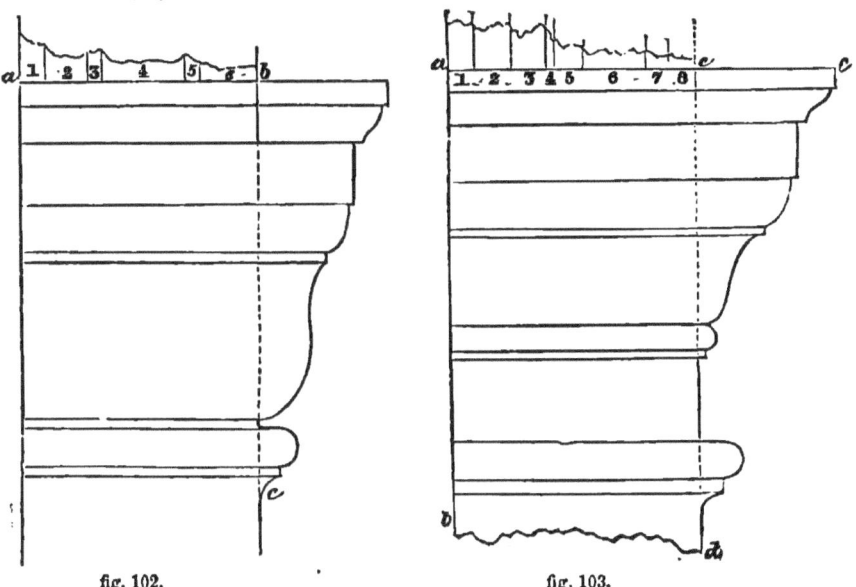

fig. 102.   fig. 103.

EXAMPLE 101, fig. 103, is the Corinthian impost. The projections

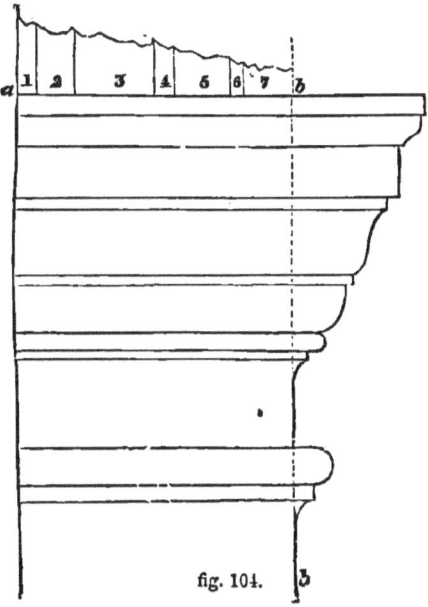

fig. 104.

being set out from the line *c d* towards *e*, the width of archivolt mouldings 1, 2, 3, &c., as *a c*, from *c* towards *a*.

EXAMPLE 102, fig. 104, is the Composite impost, the projections being set from the line *b b*. The scale from which the measurements should be taken is the same for all the imposts, being that in fig. 62.

EXAMPLE 103, fig. 105, shows a 'pediment.' c c, the tympanum, is generally filled in with sculpture. In our work on Practical Geometry, in

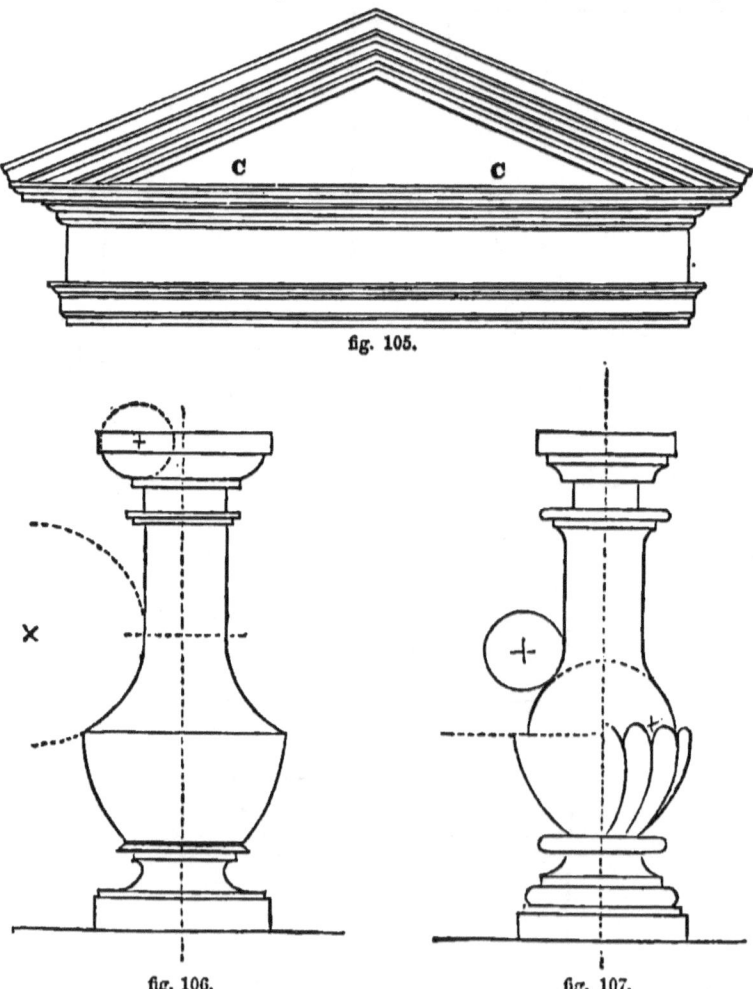

fig. 105.

fig. 106.　　　　　　　　fig. 107.

the latter part, we have shown how geometry is made applicable to the construction of the various forms of arches, vases, and balustrades. We now give, in

EXAMPLE 104, fig. 106, an elevation of the Tuscan balustrade; and in
EXAMPLE 105, fig. 107, an elevation of the Ionic.

The reader desirous of becoming acquainted with the members of the Grecian orders of architecture, and of the principles which regulate the proportions of various architectural features, of which the limits and nature of the present work do not allow us to give even a passing notice, is referred to the work previously mentioned, treating of architectural and ornamental design.

We now purpose giving examples of various architectural forms and decorations, useful to impart to the pupil a correct general idea of the method of proportioning doors, windows, &c.; and also serving as copies by which he may test his proficiency, and enable him to acquire that facility so requisite for the architectural draughtsman to possess. We shall first give forms of windows and doors.

EXAMPLE 106, fig. 108, is the elevation of an ordinary sash-window, the method of laying out of which was explained in Example 4.

fig. 108.

fig. 109.

EXAMPLE 107, fig. 109, is the elevation of a rustic window, with lozenge-shaped panes of glass. For the method of laying this out, see Example 5, fig. 5.

EXAMPLE 108, fig. 110, is an elevation of a three-light (Venetian) window, in the Italian style, drawn to a scale of one-fourth inch to the foot.

EXAMPLE 109, fig. 111, is a one-light window in the same style, to a scale of one-eighth inch to a foot.

EXAMPLE 110, fig. 112, is an elevation of a second-floor or bedroom

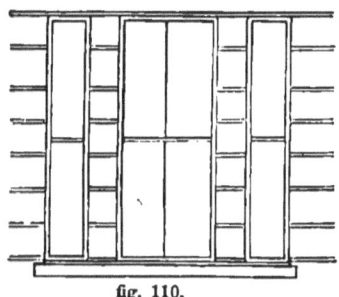
fig. 110.

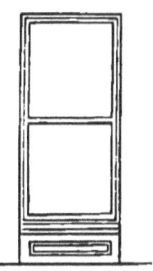
fig. 111.

window in same style, with iron ornamental balustrade in front. As a general rule, the proportion of windows should be, height twice the breadth

fig. 112.

fig. 113.

for those on the ground-floor; those on the second floor the same breadth, but of less height.

EXAMPLE 111, fig. 113, is an example of a circular-headed window, with rusticated dressings.

EXAMPLE 112, fig. 114, is the front elevation of a projecting window, of which the side elevation is given to a scale of one-quarter inch to a foot, in

EXAMPLE 113, fig. 115, which is of the same scale as the above.

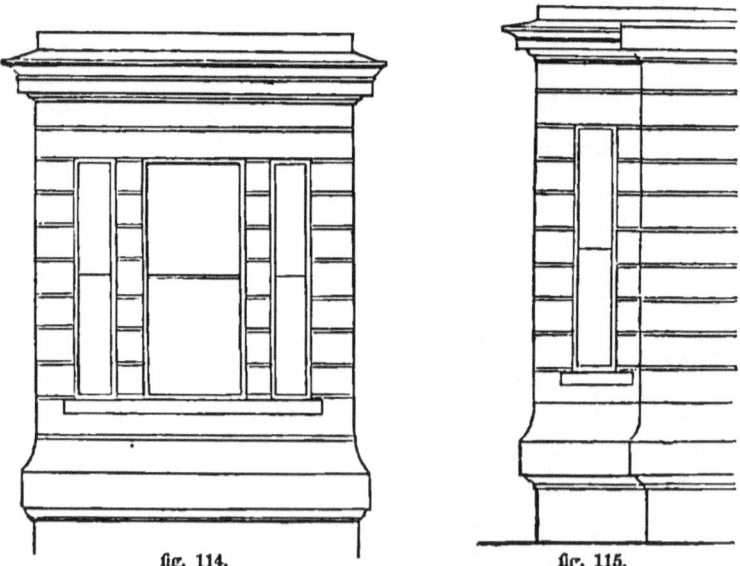

fig. 114. fig. 115.

We shall now give examples of windows placed over windows.

EXAMPLE 114, fig. 116, is the front elevation of a bay-window in the light Italian style, the plan of which shows the three sides of an octagon, with the bedroom-window over it; the scale is one-fourth of an inch to the foot. The side elevation of the bay-window is shown in

fig. 116.

EXAMPLE 115, fig. 117, which is drawn to the same scale as above.

EXAMPLE 116, fig. 118, is the elevation of a bay-window on the ground-floor, in the Domestic Gothic style, with bedroom-window over it. The scale is one-eighth of an inch to the foot.

EXAMPLE 117, fig. 119, is another sketch, showing elevation of bay-window in Italian style, with bedroom-window over; same scale as above.

EXAMPLE 118, fig. 120, shows the elevation of window over window in the Tudor style; scale three-sixteenths of an inch to the foot.

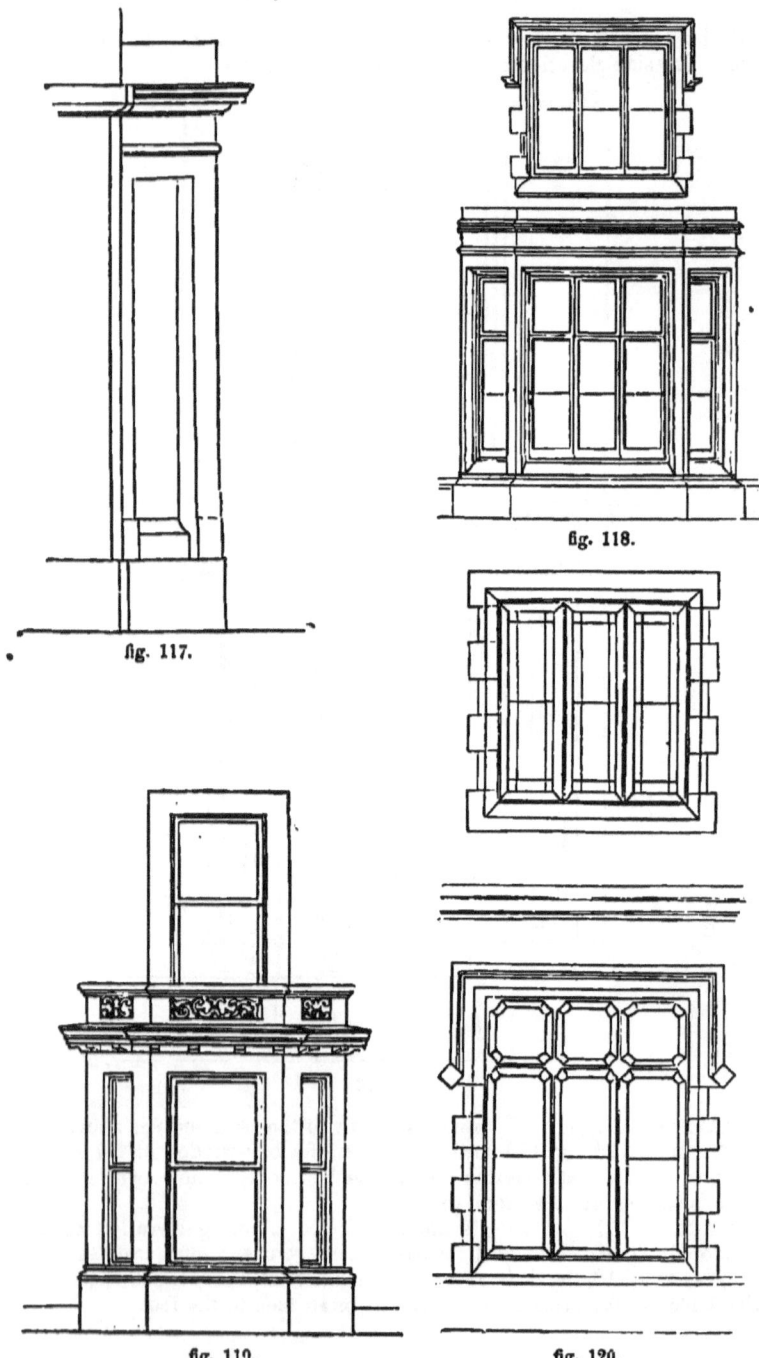

fig. 117.

fig. 118.

fig. 119.

fig. 120.

EXAMPLE 119, fig. 121, is the front elevation of a bay-window on ground-floor, with projecting or oriel window over it on bedroom-floor, in the Elizabethan or Jacobin style, drawn to a scale of one eighth inch to a foot. The side elevation of this drawing is shown in

EXAMPLE 120, fig. 122, same scale as above.

EXAMPLE 121, fig. 123, is a sketch, showing front elevation of Venetian or three-light window on ground-floor in Italian style, with bedroom-window over, with ornamental dressings and segmental pediment.

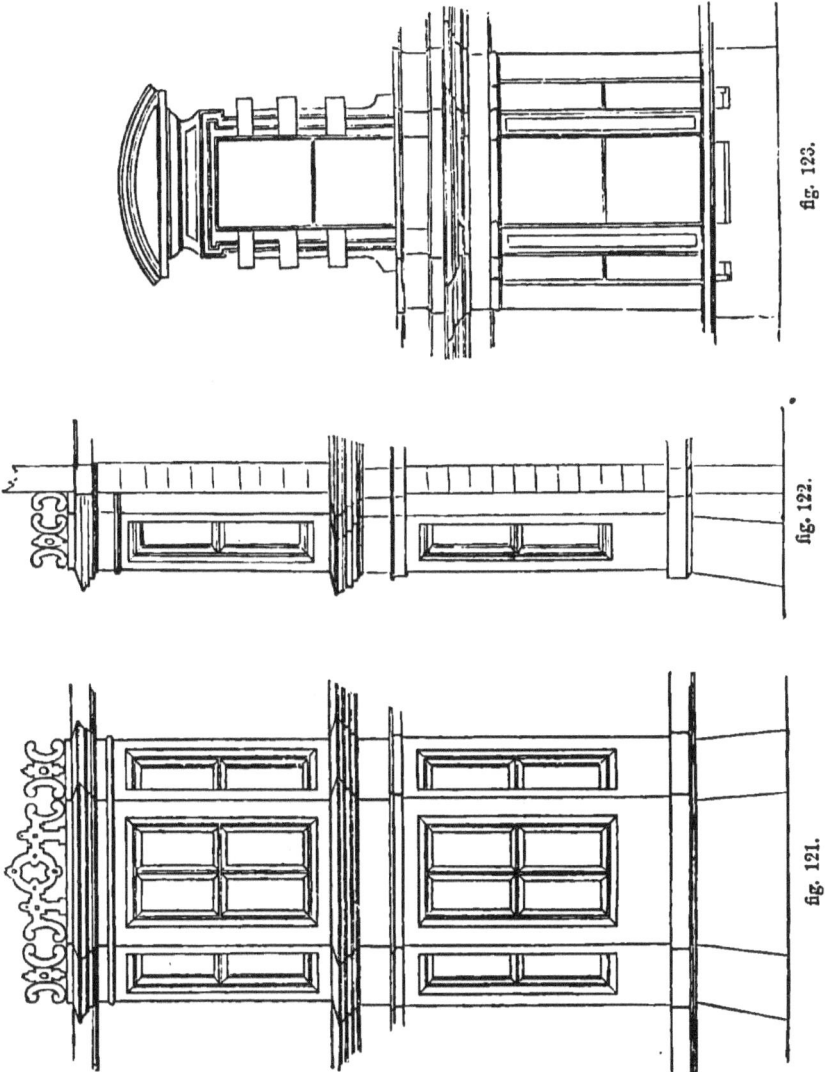

We now proceed to give examples of doors, and windows over doors. First, as to doors, of which, in

EXAMPLE 122, fig. 124, we give the elevation of one in the Roman style.

EXAMPLE 123, fig. 125, is the front elevation of another form in the Italian style.

fig. 124.

fig. 125.

fig. 126.

EXAMPLE 124, fig. 126, is the elevation of a form suitable for a public building, with vermiculated dressings. Another form is given in

EXAMPLE 125, fig. 127.

EXAMPLE 126, fig. 128, is front elevation of door, with vermiculated dressings, in the Italian style (of which fig. 111 is the window belonging to same design), with circular-headed window over it. The scale is one-quarter inch to the foot.

EXAMPLE 127, fig. 129, is front elevation of door, to the house of which fig. 114 is the principal window. The scale is one-quarter inch.

fig. 127.

fig. 128.

fig. 129.

EXAMPLE 128, fig. 130, is front elevation of door at the end of house, with window on second floor over it. This example is in the same style as fig. 123, which is the principal window to same house of which this figure is the door. The scale is one-eighth inch.

EXAMPLE 129, fig. 131, is front elevation of doorway to house of which fig. 116 is the window, having over it a circular-headed window in bedroom-floor. The scale is one-quarter inch. The side elevation of the door in this drawing is given in

EXAMPLE 130, fig. 132, the scale of which is the same as above.

EXAMPLE 131, fig. 133, is front elevation of principal door to a house, with bedroom-window over it, with ornamented dressings. Scale one-eighth inch.

The central portion of house (of which figs. 123 and 130 are parts of the same design).

fig. 131.

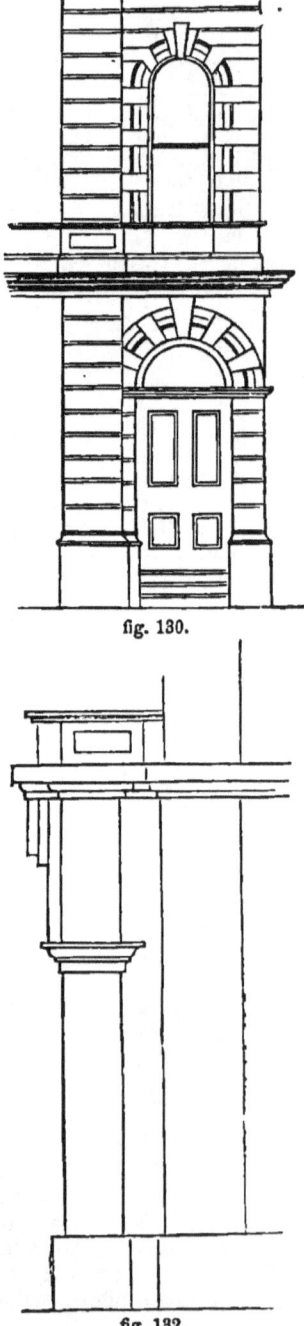

fig. 130.

fig. 132.

EXAMPLE 132, fig. 134, is front elevation of principal door to house, in Domestic Gothic style (of which the drawing in fig. 118 is the window), with closet-window over it on bedroom-floor.

fig. 133.

fig. 134.

fig. 135.

EXAMPLE 133, fig. 135, is elevation of principal entrance to house, in Tudor style, of which the drawing in fig. 120 is the window.

ENGINEERING, AND MECHANICAL DRAWING-BOOK. 67

EXAMPLE 134, fig. 136, is elevation of principal entrance, with window over it, of house of which fig. 121 is the window.

We now give the elevations of a few examples of fireplaces, the first of which is in the Tudor style, and is shown in
EXAMPLE 135, fig. 137. The part c shows the profile of the mouldings. Another form, in same style, is given in
EXAMPLE 136, fig. 138.
EXAMPLE 137, fig. 139, is in the Italian style; A shows the profile of

fig. 136.

68   ILLUSTRATED ARCHITECTURAL,

the skirting-board running round the room, of which the lines at B show the front elevation.

EXAMPLE 138, fig. 140, is in the Elizabethan style. In these examples

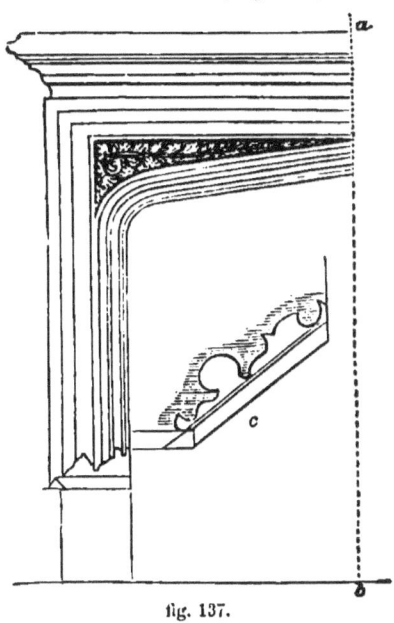

fig. 137.

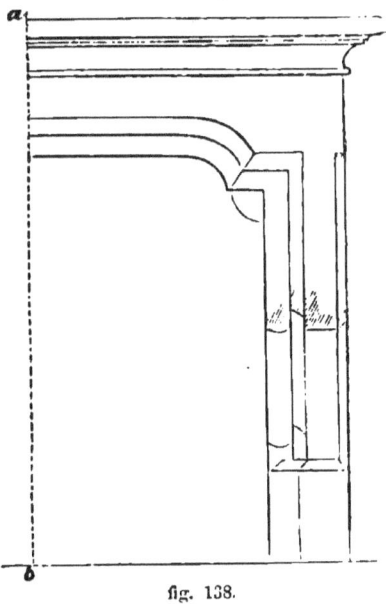

fig. 138.

of fireplaces, we have only shown half, the other being an exact counterpart. The pupil should, however, draw them complete, the line *a b* being the centre-line.

fig. 139.

fig. 140.

ENGINEERING, AND MECHANICAL DRAWING-BOOK.

We now proceed to the more elaborate copies, in which the pupil will find ample exercise for the display of that facility for copying which the foregoing lessons have been designed to impart. From the limits of the page we have been compelled to adopt a small scale; it is to be understood, however, that the pupil is to copy them to a larger one, at least twice as large as those we have adopted.

EXAMPLE 139, fig. 141, is the front elevation of a school-house, with railings to the front. Another design is given in

EXAMPLE 140, fig. 142.

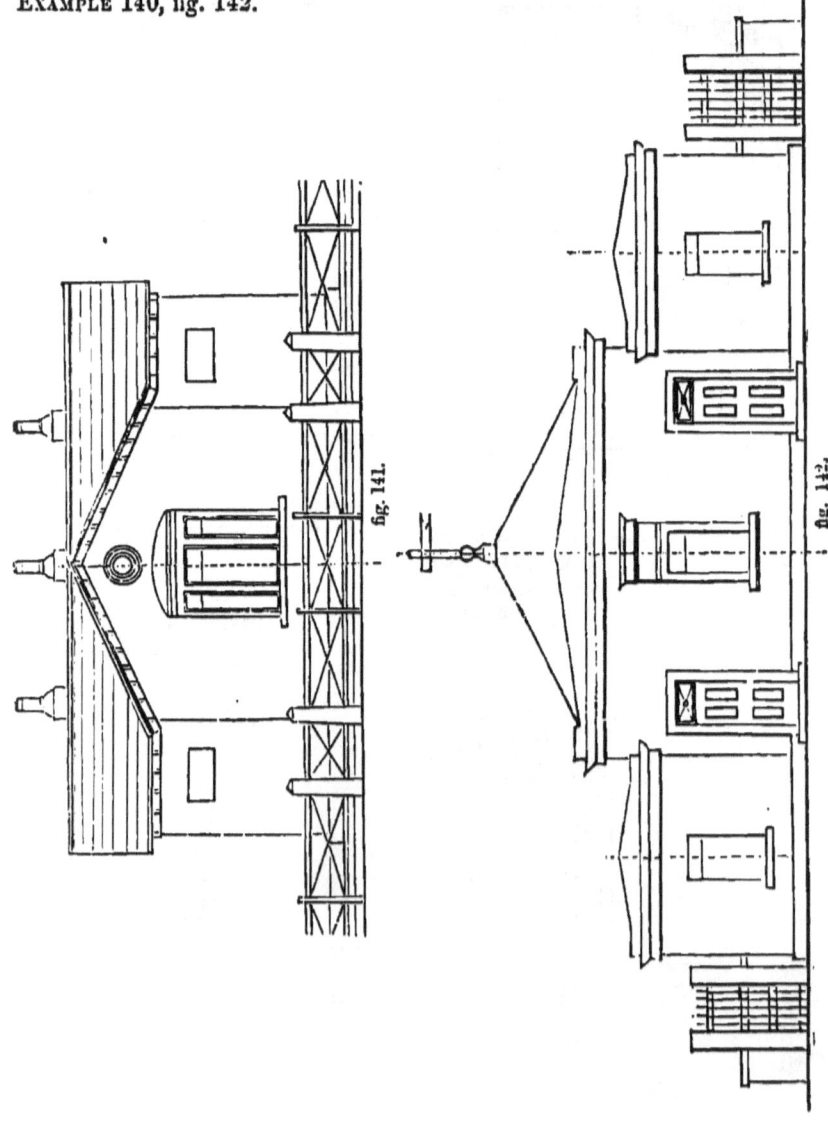

EXAMPLE 141, fig. 143, is the front elevation of a row of cottages, drawn to a scale of one-eighth inch to the foot.

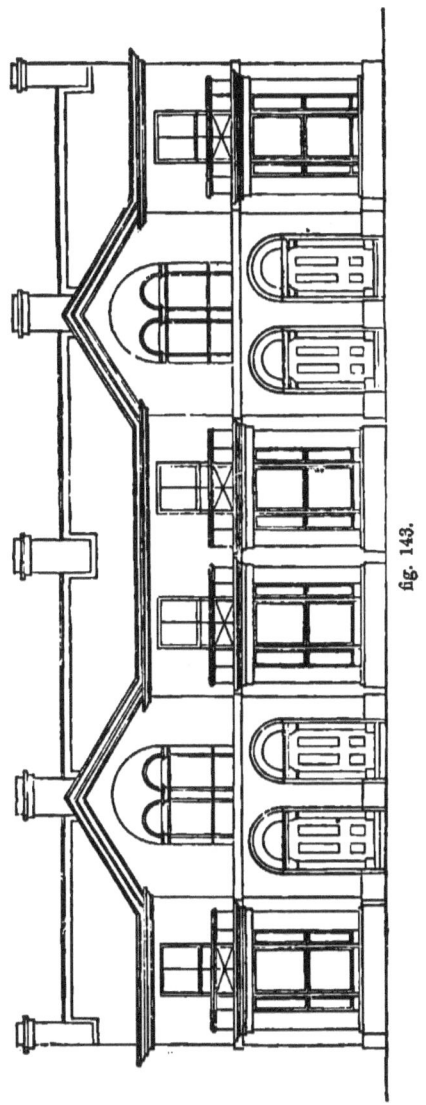

fig. 143.

EXAMPLE 142, fig. 144, is the front elevation of a shop-front, drawn to a scale of one-eighth inch to the foot.

fig. 144.

EXAMPLE 143, fig. 145, is front elevation of a greenhouse, of which the end elevation is given in

EXAMPLE 144, fig. 146, and the plan in

EXAMPLE 145, fig. 147; they are all drawn to a scale of one-sixteenth inch to the foot.

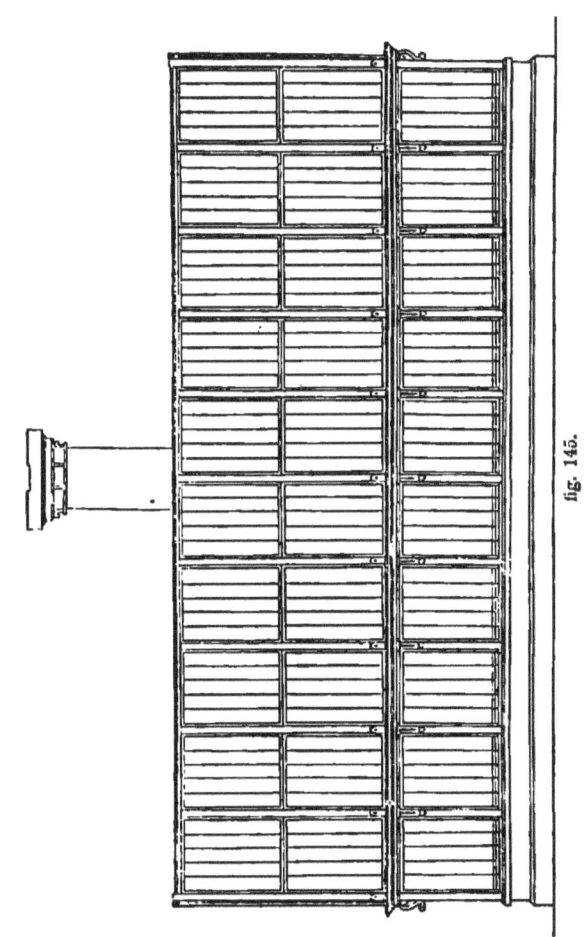

fig. 145.

In order to give the pupil an idea as to the way in which a set of plans are set out for the guidance of the artisan and workman, we have prepared a series of drawings illustrative of the design for a town-house in the Italian style. It is necessary to mention that the design when finished is double that given in the drawings; two houses being attached, the other half of the drawing (not shown) is the exact counterpart of that given in the copy. The scale we have adopted is one-eighth inch to the foot. The pupil, in copying them, should make the scale at least double this, or one-fourth inch to the foot.

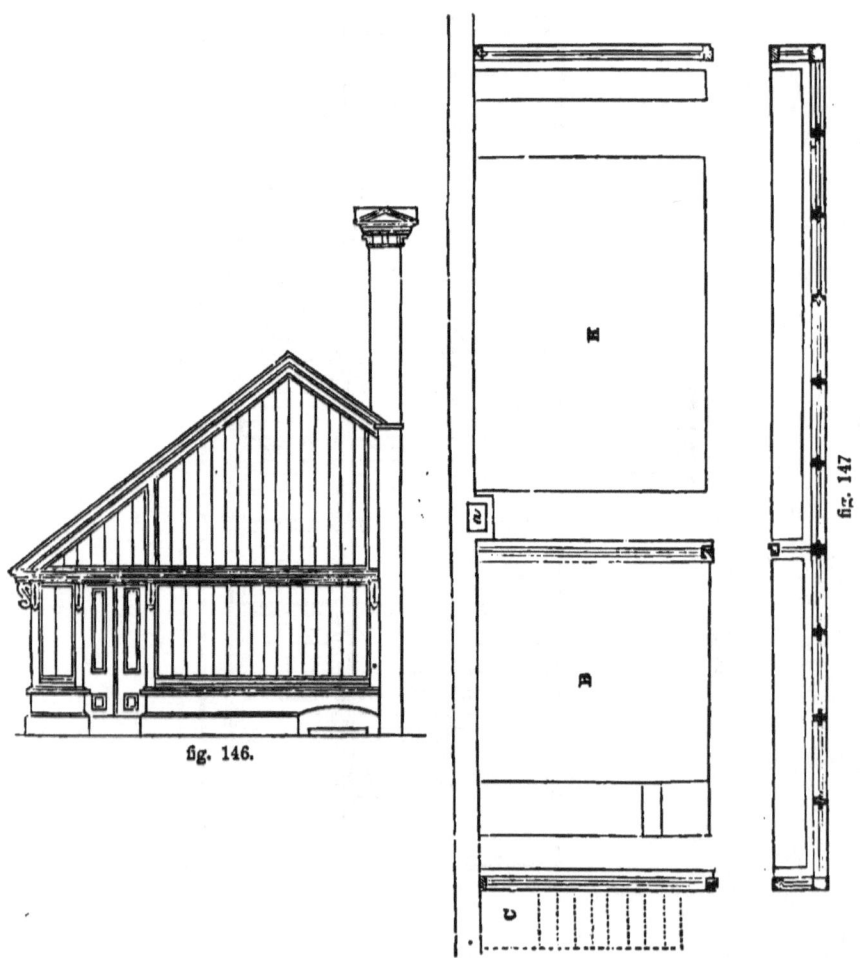

fig. 146.

fig. 147.

EXAMPLE 146, fig. 148, is the "*basement plan*" of the house; the line *a b* is the centre-line.

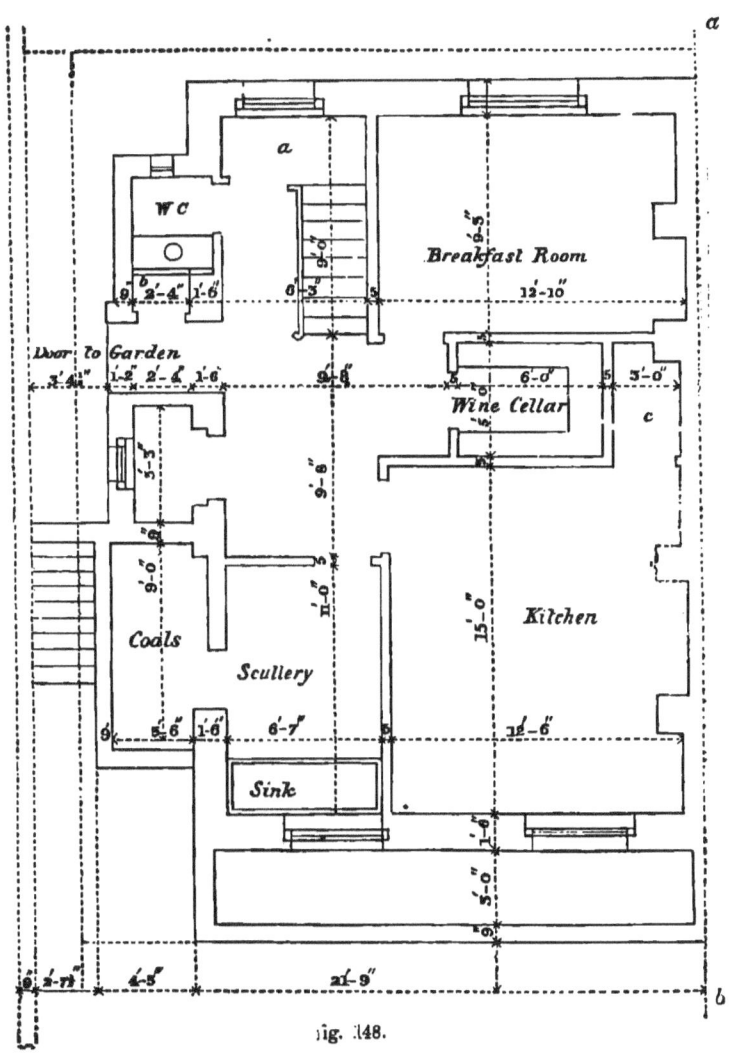

fig. 148.

ENGINEERING, AND MECHANICAL DRAWING-BOOK.   75

EXAMPLE 147, fig. 149, is the "*ground plan.*"

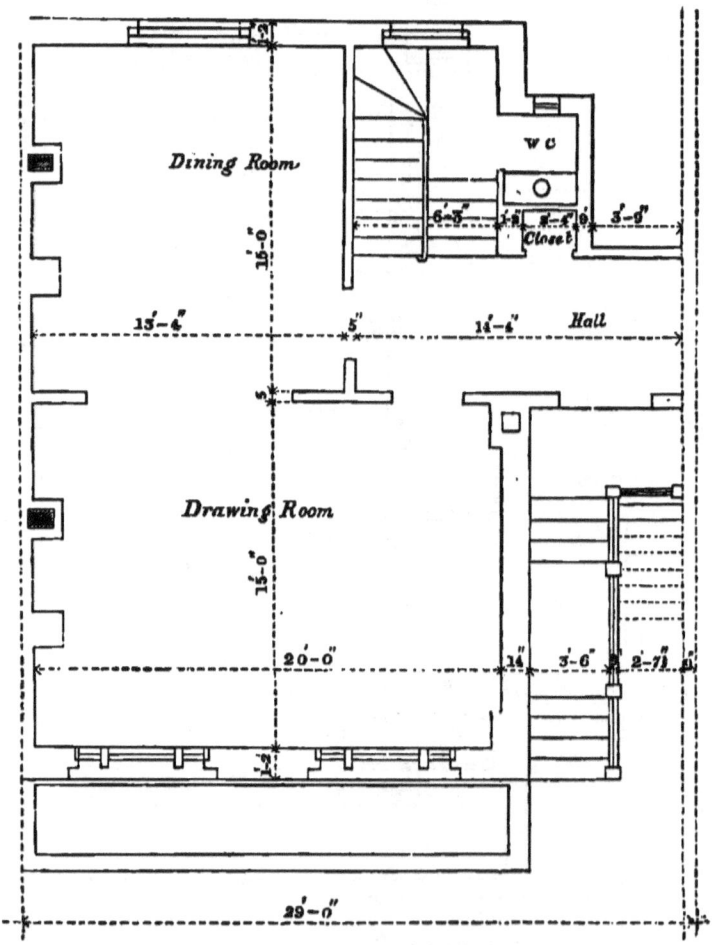

fig. 149.

76 ILLUSTRATED ARCHITECTURAL,

EXAMPLE 148, fig. 150, is half plan of *first bedroom floor*.

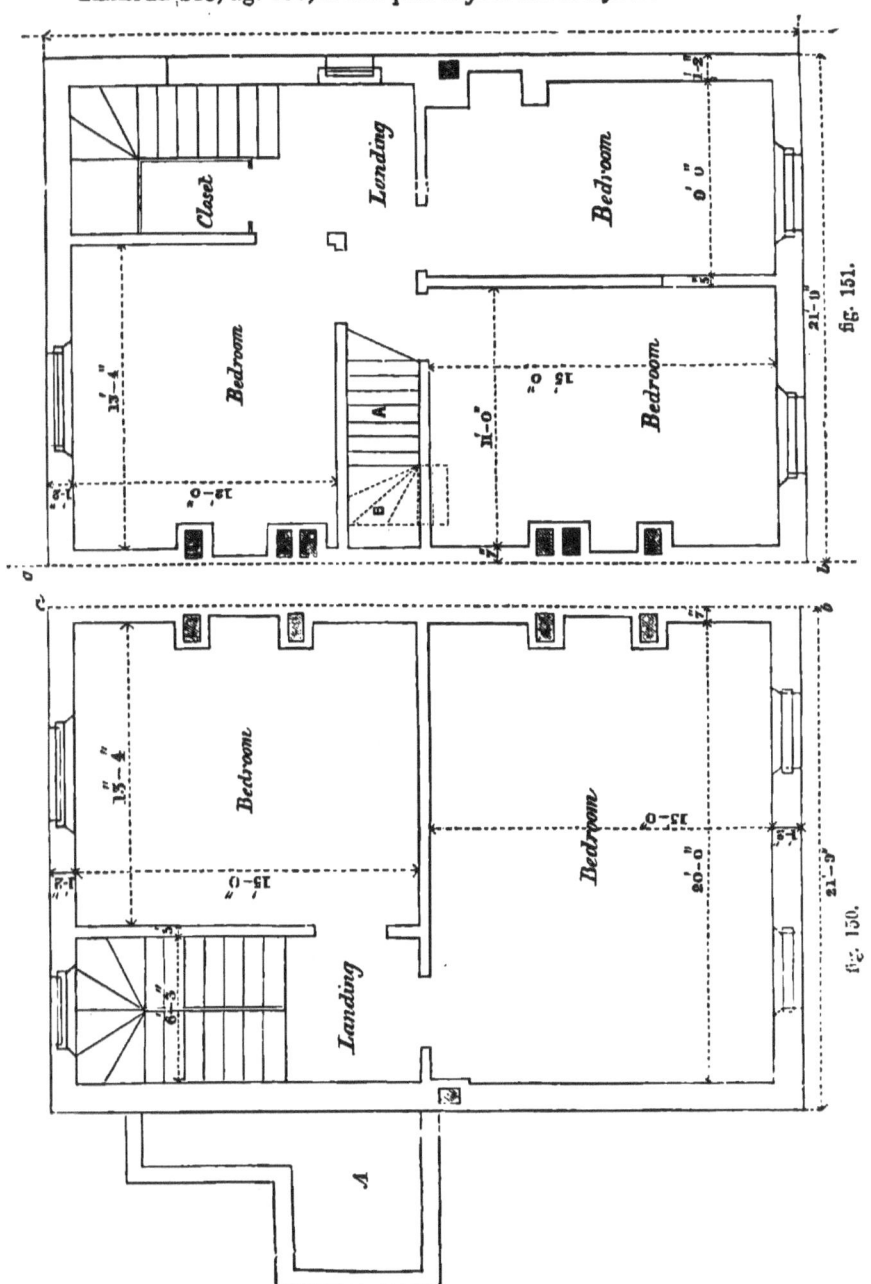

ENGINEERING, AND MECHANICAL DRAWING-BOOK. 77

EXAMPLE 149, fig. 151, is the half plan of *second bedroom floor*.

EXAMPLE 150, fig. 152, is half *front elevation*. From the minuteness of the scale we give detail drawings, which will show the decorative portions more fully than in the sketch. The first of these we give is the elevation

fig. 152.

78 ILLUSTRATED ARCHITECTURAL,

of the first bedroom-floor window, and its section drawn to a larger scale; it is shown in

EXAMPLE 151, fig. 153. The front elevation of cornice is given in EXAMPLE 152, fig. 154; and the section showing form of bracket in

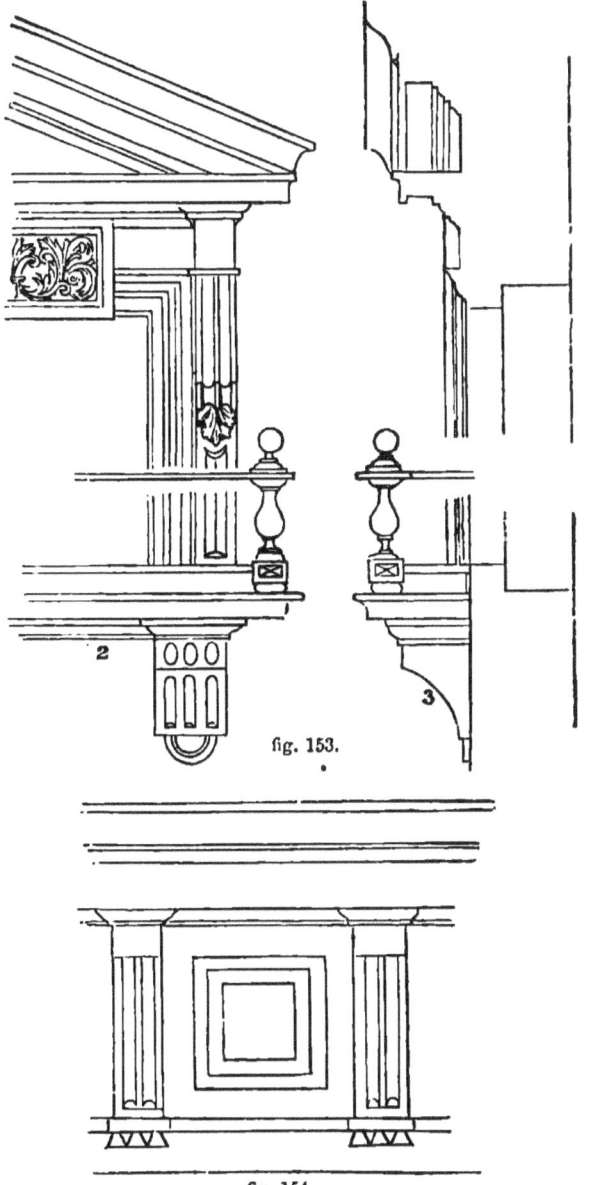

fig. 153.

fig. 154.

ENGINEERING, AND MECHANICAL DRAWING-BOOK. 73

EXAMPLE 153, fig. 155. The elevation of chimney is shown in
EXAMPLE 154, fig. 156, and the elevation of cornice and finial to principal entrance is given in
EXAMPLE 155, fig. 157.

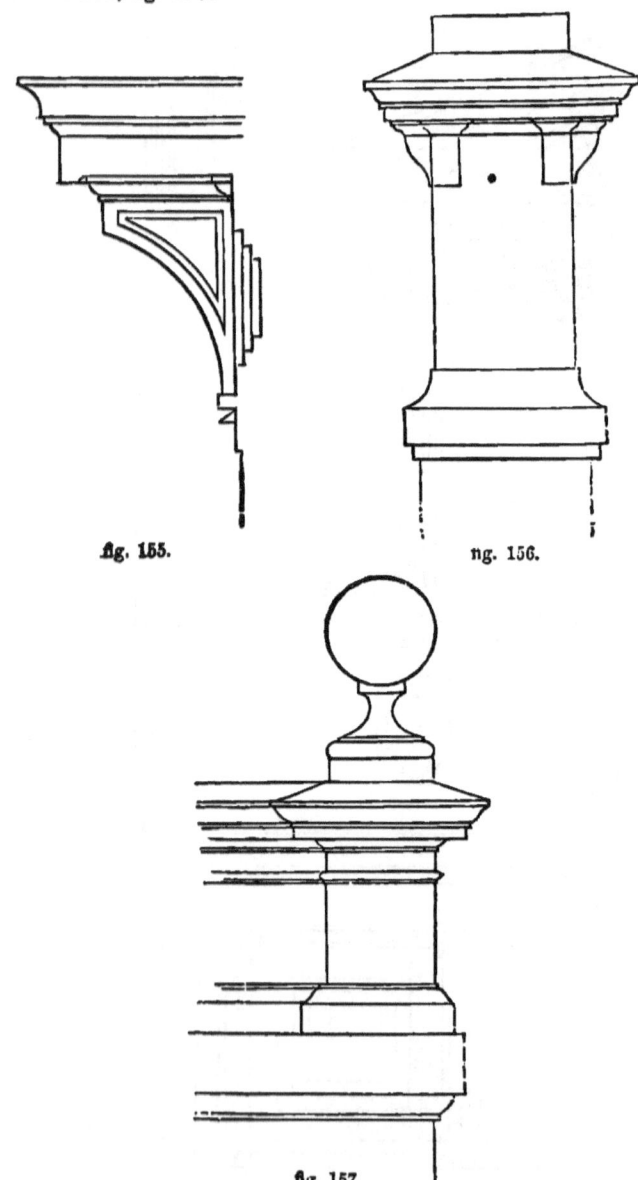

fig. 155.

fig. 156.

fig. 157.

80   ILLUSTRATED ARCHITECTURAL,

EXAMPLE 156, fig. 158, is *end elevation*. We give this in full, as one side is different from the other. The *half back elevation* is given in

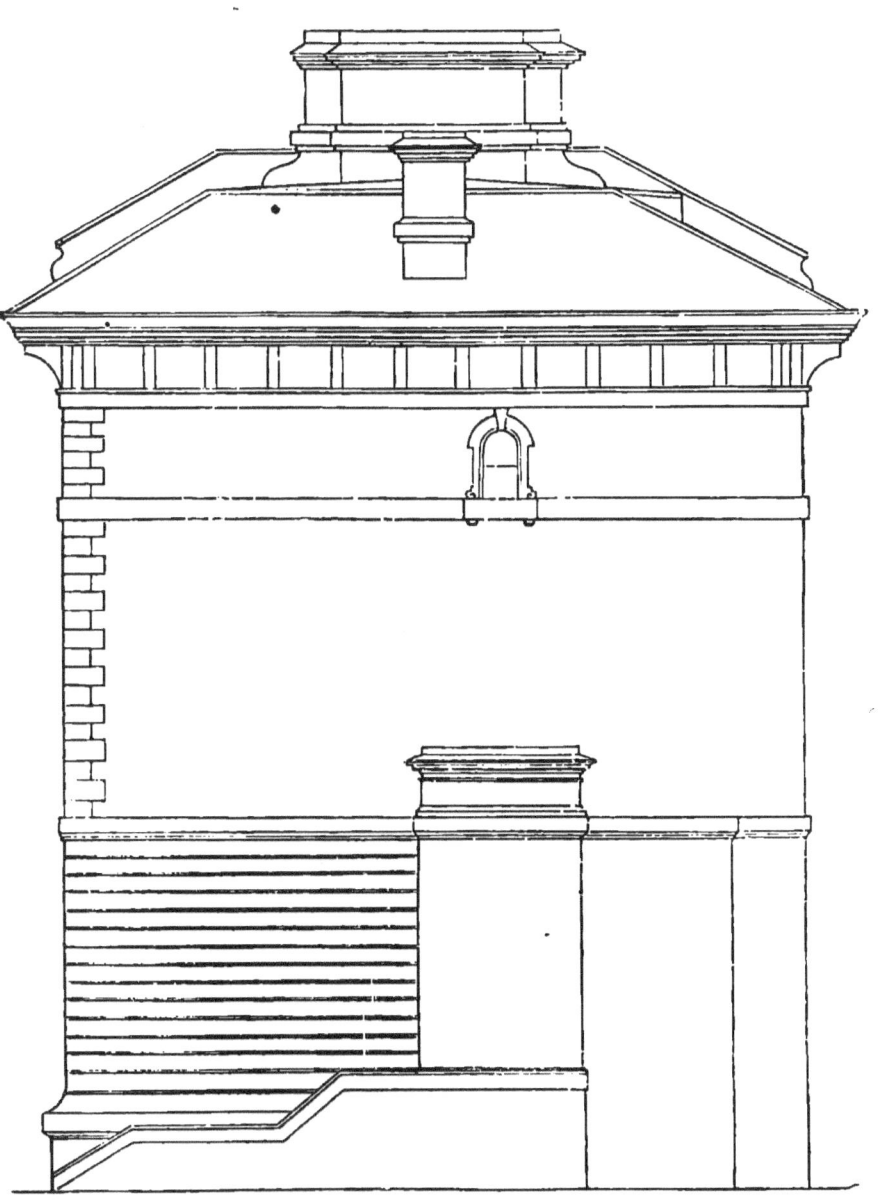

fig. 158.

EXAMPLE 157, fig. 159. The *transverse section* is taken across the *plan*. The right-hand half of this is given in

fig. 159.

EXAMPLE 158, fig. 160; the left-hand half in
EXAMPLE 159, fig. 161. The same letters of reference apply to both drawings. The pupil should make this section in one complete drawing. We have only shown one part up to the roof-line, the other without the chimney-shaft, but showing the roof-timbers. The pupil should be able to finish these sections from the other drawings.

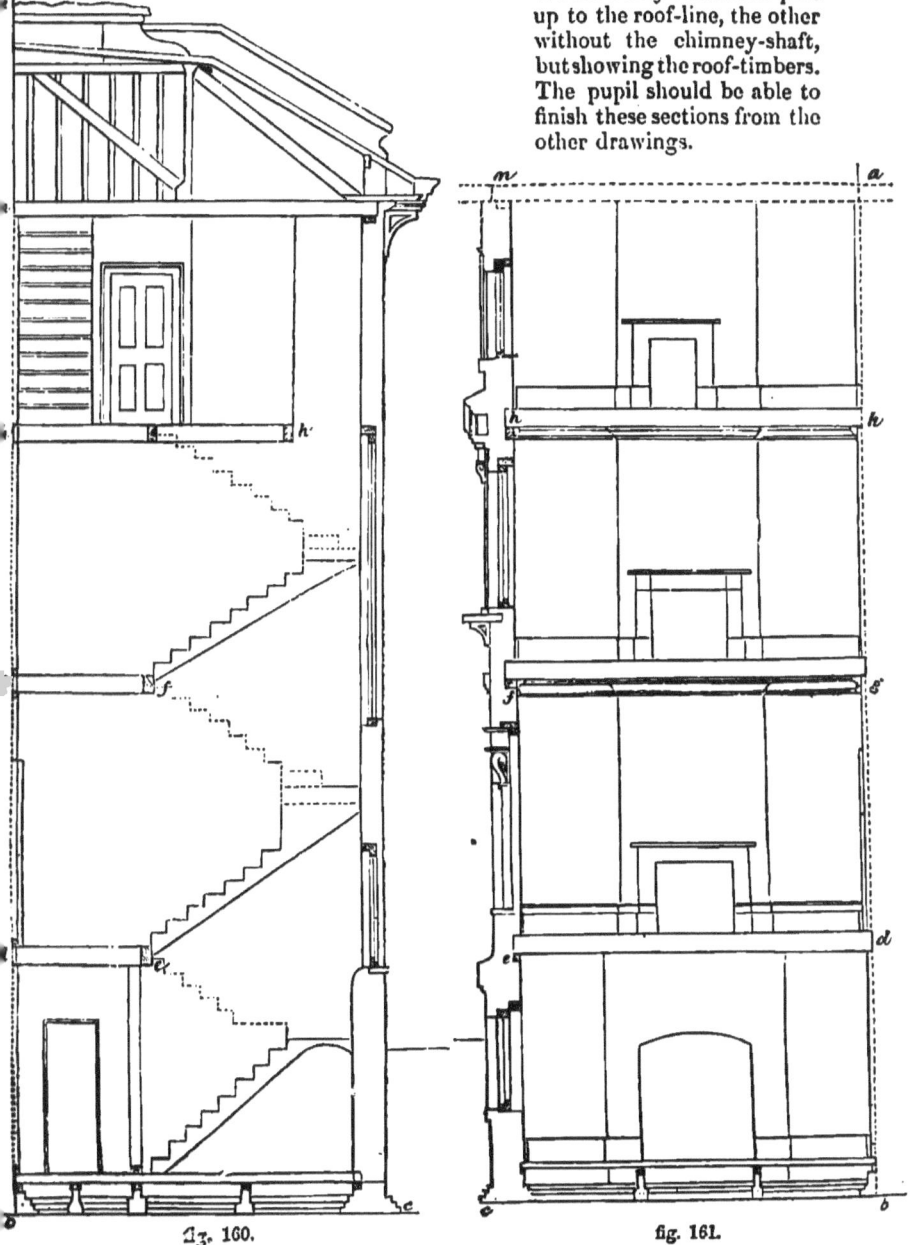

fig. 160.   fig. 161.

ENGINEERING, AND MECHANICAL DRAWING-BOOK. 83

EXAMPLE 160, fig. 162, is half plan of roof, showing timbers. The other half, showing the slated surface, and position of flues, is given in

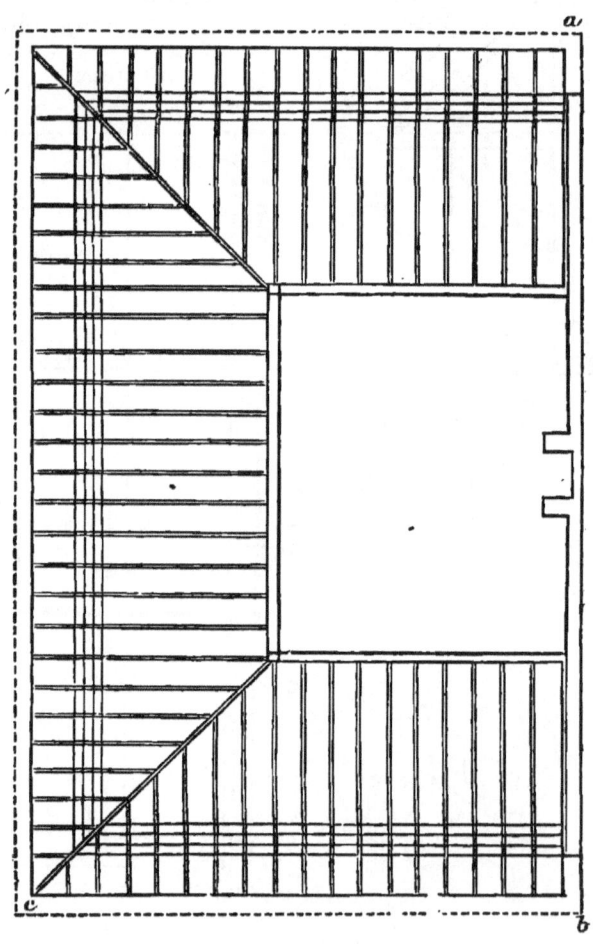

fig. 162.

EXAMPLE 161, fig. 163. In setting out this, the pupil may copy it, by drawing the line *a b*, and continuing it to *c;* measuring from *d* to *c* will give the position of the end *a b* of the flue. From *d* to *f* the position of the point *e, g h*, the distance of line *e g n* from line *d h*.

EXAMPLE 162, fig. 164, is a transverse section of a fireproof vaulted warehouse, where *a, a* are the retaining walls, a strong iron tie passing

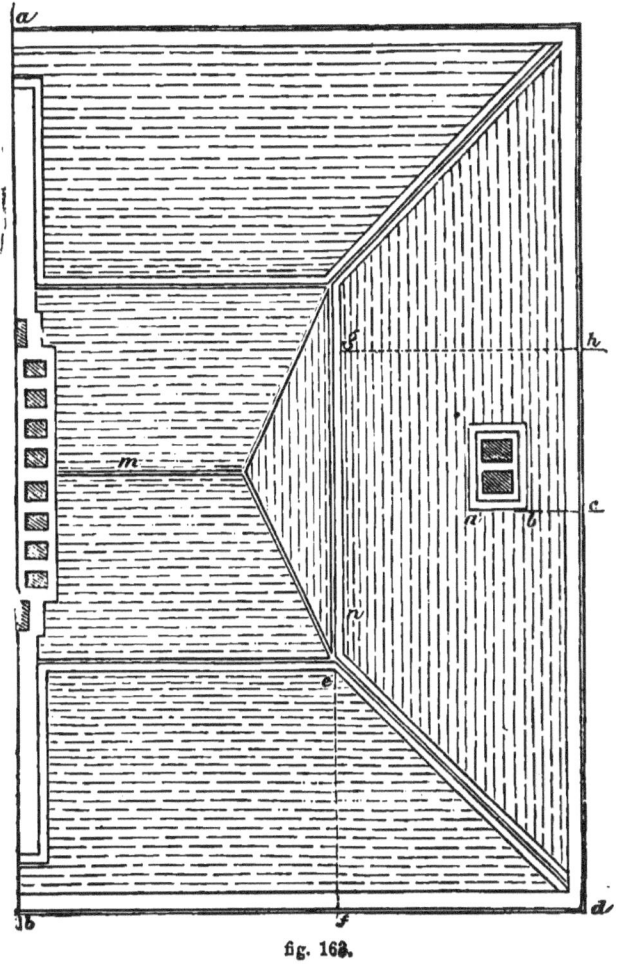

fig. 163.

through both, and secured by a screw bolt, and nut. The arches m, m are described from their centres g, g on the lines h, h, springing from the pillars c, d: the arch n is described from centre i.

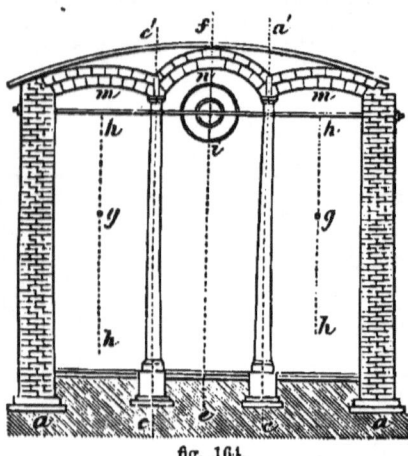

fig. 164.

EXAMPLE 163, fig. 165, is a transverse section of a fireproof cottage.

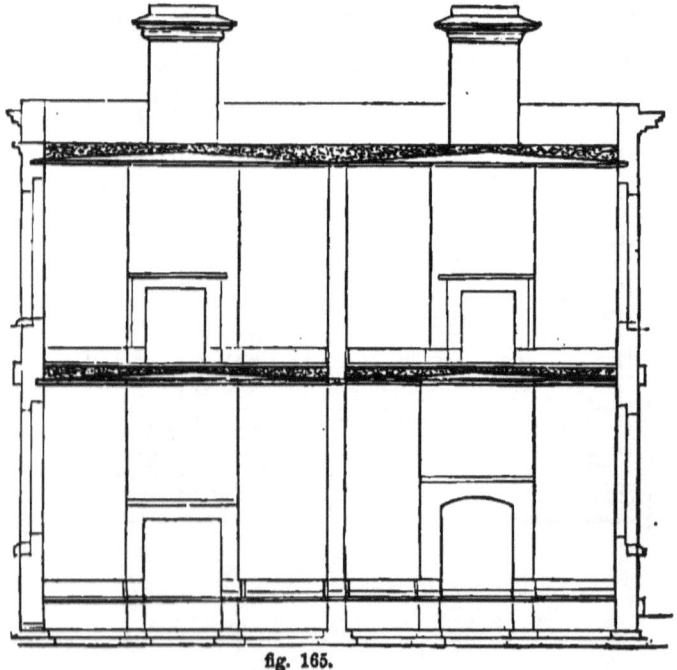

fig. 165.

In our work in the present series, the *Illustrated Drawing-Book*, we have given directions for delineating architectural subjects perspectively. We now present a few additional examples, which will serve as copies with which the pupil may still further exercise himself in architectural drawing; premising that in this department he is supposed to have the advantage of a knowledge of the rules by which objects are put in perspective, and a facility in copying such subjects as depend chiefly on the eye, aided by a readiness of hand in pencilling. These desiderata are indispensable before the pupil can copy the examples which we are now to present to his notice; for assistance as to the readiest means of attaining them, we beg to refer the pupil to the above work.

EXAMPLE 164, fig. 166, is the perspective drawing of a public asylum, in the Italian style, with a campanile tower.

fig. 166.

fig. 167.

fig. 168.

EXAMPLE 165, fig. 167, is a perspective sketch of the interior of an apartment, with carved panels, &c., in the Italian style.

We now present a few examples of churches perspectively delineated; the first of these,

EXAMPLE 166, fig. 168, is a perspective drawing of a church in the Early-English style.

fig. 169.

EXAMPLE 167, fig. 169, which is in the Early-Decorated or Pure Geometrical style. The peculiarities of the various styles of Gothic architecture will be seen by an inspection of figs. 202, 203, &c.

EXAMPLE 168, fig. 170, is in the Transitional from Decorated to Perpendicular.

EXAMPLE 169, fig. 171, is in the Middle or Second Pointed Period.

fig. 170.

fig. 171.

EXAMPLE 170, fig. 172, is in the Early-Decorated style.

fig. 172

EXAMPLE 171, fig. 173, is in the Early-English style.

fig. 173.

EXAMPLE 172, fig. 174, represents in perspective the interior of part of a church (the nave) in the Norman style. This is considered to be a fine specimen of the architecture of the period.

fig 174.

ENGINEERING, AND MECHANICAL DRAWING-BOOK. 93

EXAMPLE 173, fig. 175, represents the interior of the Lady chapel in Tynemouth Priory church; the architectural features of which belong somewhat both to the Decorated and Perpendicular styles.

fig. 175.

We now proceed to give a few illustrations of architectural ornament; the drawings of which are nearly in all the instances produced by hand, only here and there aided by the drawing-board and instruments. A knowledge of pencil-sketching is therefore necessary for these examples.

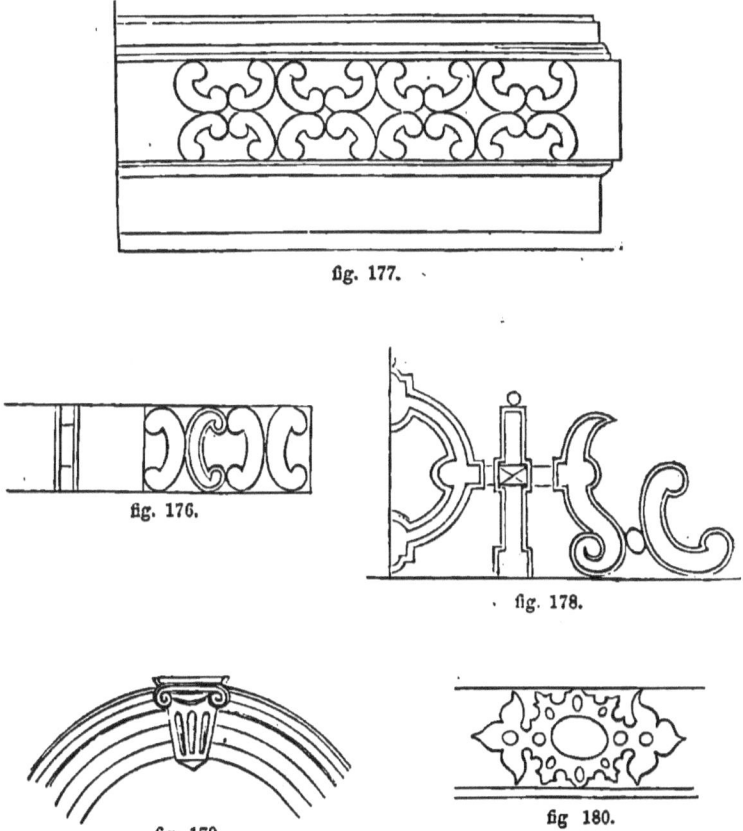

fig. 177.

fig. 176.

fig. 178.

fig. 179.

fig 180.

EXAMPLE 174, fig. 176, is the elevation and end view of a pierced parapet in the Elizabethan style.

EXAMPLE 175, fig. 177, is a side elevation of panelling, in the same style as the last figure.

EXAMPLE 176, fig. 178, is another example of a pierced parapet, in the same style as in fig 177.

EXAMPLE 177, fig. 179, is the front elevation of a key-stone.

EXAMPLE 178, fig. 180, is another example of raised panel, in the same style as fig. 177.

fig. 181.

fig. 182

fig. 183.

fig. 184.

EXAMPLE 179, fig. 181, is a design for a Gothic panel.

EXAMPLE 180, fig. 182, is the Grecian ornament known as the "honeysuckle."

EXAMPLE 181, fig. 183, is part of an ornamented frieze for the Ionic column.

EXAMPLE 182, fig. 184, is an ornament sometimes used in filling up the space called "metopes" in the Doric order. (See p. 42, ex. 71.)

EXAMPLE 183, fig. 185, is a design for a frieze and cornice.

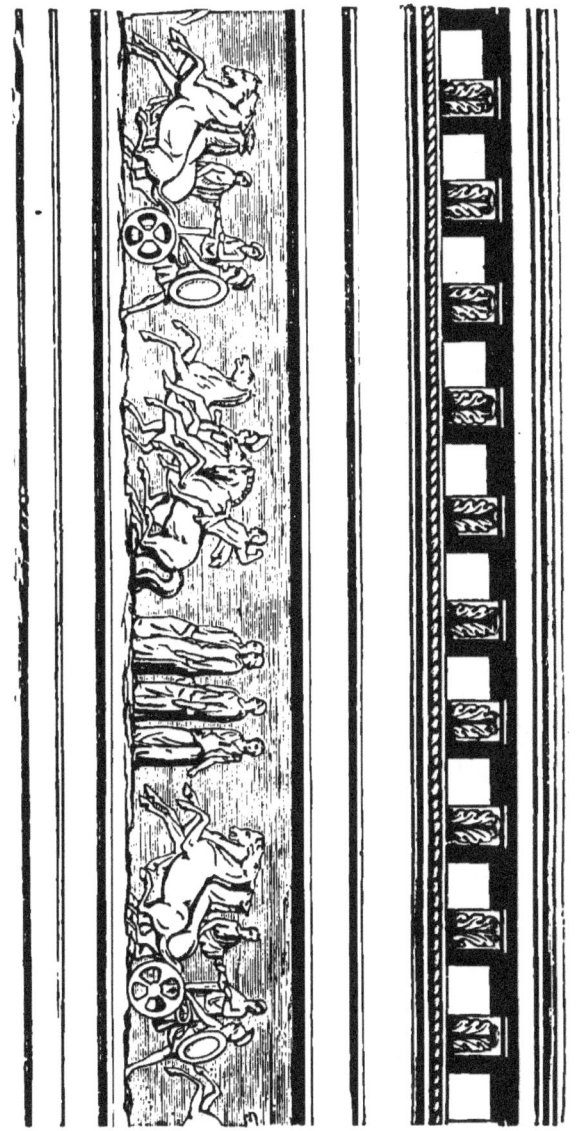

fig. 185.

EXAMPLE 184, fig. 186, is the elevation of a sculptured pilaster forming part of a chimney-jamb.

ENGINEERING, AND MECHANICAL DRAWING-BOOK.   97

EXAMPLE 185, fig. 187, is a form of ornament sometimes used in place of balustrades.

EXAMPLE 186, fig. 188, is an example of bracket, of which the side view is given in

EXAMPLE 187, fig. 189.

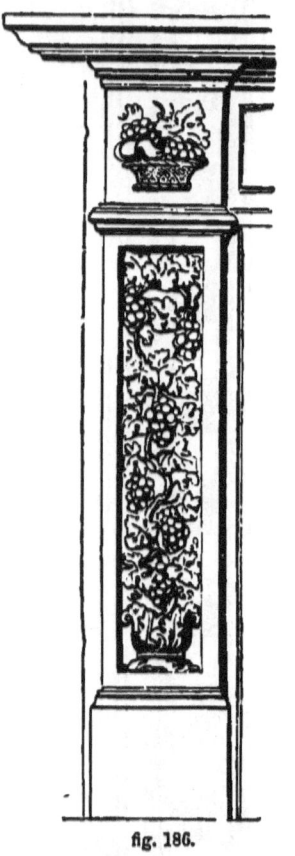

fig. 186.

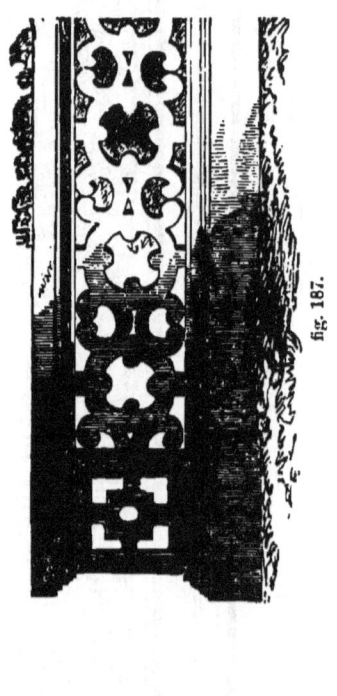

fig. 187.

EXAMPLE 188, fig. 190, is a perspective view of a Grecian 'scroll truss.'

EXAMPLE 189, fig. 191, is an elevation of an Elizabethan scroll truss.

EXAMPLE 190, fig. 192, is an exemplification of the ornament called the 'fret.' Another form is given in

EXAMPLE 191, fig. 193.

EXAMPLE 192, fig. 194, is an exemplification of the ornament termed the 'guilloche.' Another example is given in

EXAMPLE 193, fig. 195.

G

98 ILLUSTRATED ARCHITECTURAL,

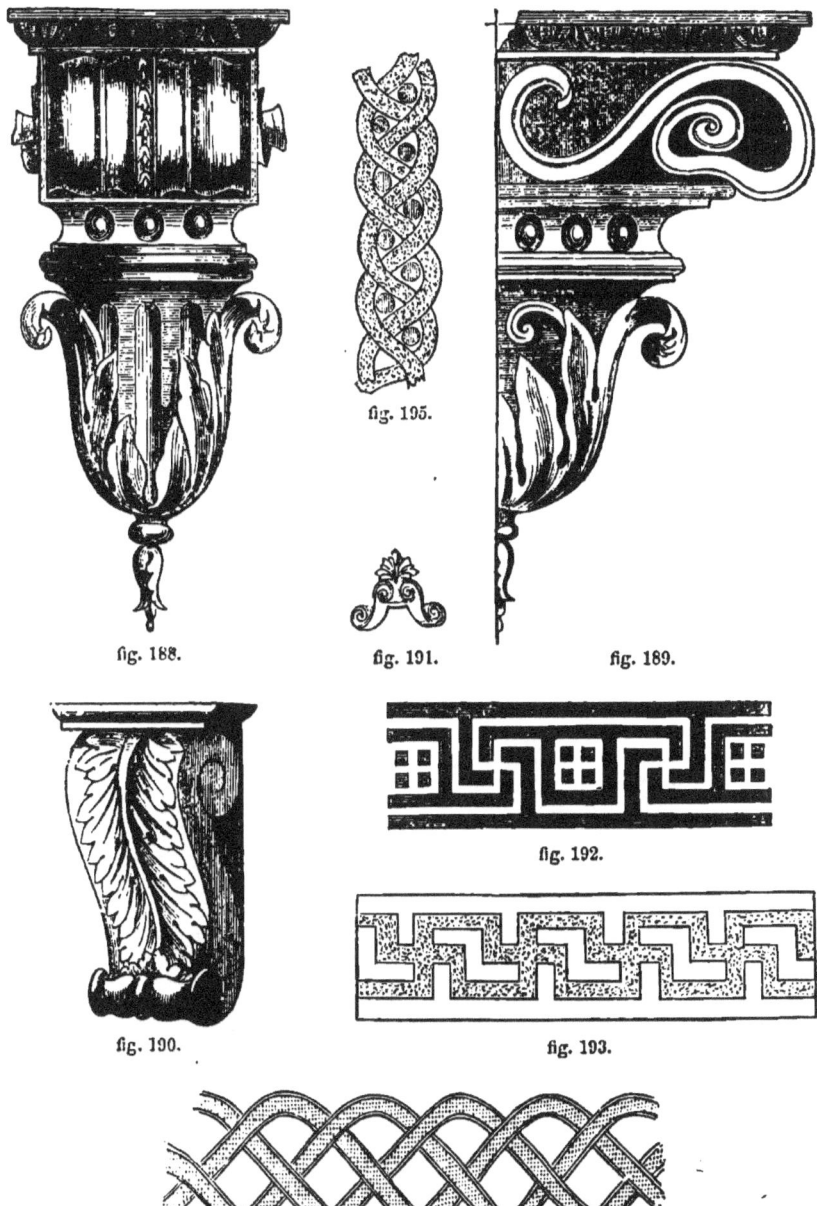

fig. 188.   fig. 195.   fig. 189.

fig. 191.

fig. 190.   fig. 192.

fig. 193.

fig. 194.

In the work on *Practical Geometry* we have given examples of outlines of vases, with the methods of describing their curves. We now present a specimen with the outlines ornamented.

EXAMPLE 194, fig. 196.

EXAMPLE 195, fig. 197, is an example of 'vase and pedestal.'

fig. 196.

fig 197

100 ILLUSTRATED ARCHITECTURAL,

EXAMPLE 196, fig. 198. Another example of vase, with the outlines ornamented.

EXAMPLE 197, fig. 199. Design for a Gothic monument.

fig. 198.

fig. 199.

ENGINEERING, AND MECHANICAL DRAWING-BOOK. 101

EXAMPLE 198, fig. 200. A design for a fountain.
EXAMPLE 199, fig. 201, is the elevation of a stained window in the geometrical style.

102 ILLUSTRATED ARCHITECTURAL,

We now, as concluding this department of our treatise, proceed to give a series of designs, exemplifying by inspection the peculiarities of the various periods of Gothic architecture as generally received.

EXAMPLE 200, fig. 202, is an elevation of a Norman window.

fig. 202.

EXAMPLE 201, fig. 203, is the Early-English (or Lancet). This style

fig. 203.

ENGINEERING, AND MECHANICAL DRAWING-BOOK.   103

succeeded the Norman, and was followed by the Decorated, the tracery of which was distinguished by geometrical lines, as in

EXAMPLE 202, fig. 204; and in the later instances by flowing lines, termed curvilinear, as in

fig. 204.

EXAMPLE 203, fig. 205. The Perpendicular is derived from the Deco

fig 205.

rated; its distinguishing feature is the perpendicular lines of the tracery, as seen in

EXAMPLE 204, fig. 206.

For further information on the styles and peculiarities of Gothic Architecture, see the work on " Ornamental and Architectural Design."

fig. 206.

SECTION II.

ENGINEERING DRAWING.

In this section we purpose explaining, chiefly by appropriate illustrations, the methods of delineating those subjects which are found more particularly the branches of what is generally designated as Civil Engineering, whether these be shown in plans, maps, elevations, or sections. As the rules, or more properly the methods, to be observed in copying subjects of pure outline, where the drawing-board and instruments are available, will obviously be very similar to those which we have already detailed in the First Section, we do not consider it necessary to multiply examples of outlines, such as bridges, &c. The pupil desirous of studying Civil Engineering as a profession will find numerous examples which may serve as "copies" in the more technical and strictly professional works which it will be his duty to consult. We shall content ourselves with giving one or two examples of the method of setting out copies of bridges, &c.

EXAMPLE 1, fig. 1. Bisect any two of the piers, as $a\,b$, $c\,d$, in the points a and c. Draw lines $a\,m$, $c\,d$; put in the piers; divide $a\,c$ into two equal

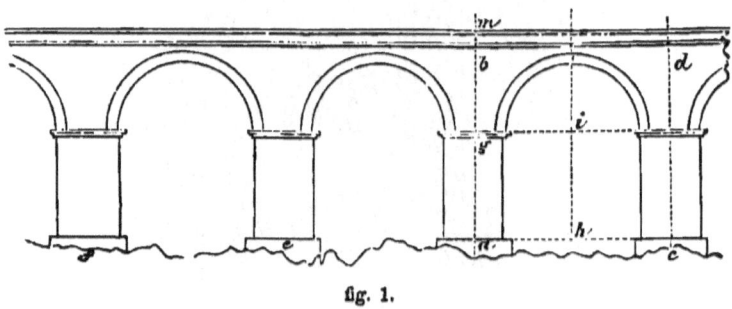

fig. 1.

portions at the point h; parallel to $c\,d$ draw $h\,i$; mersure to i. This will be the centre of the arch. In like manner the aqueduct arches in

106 ILLUSTRATED ARCHITECTURAL,

EXAMPLE 2, fig. 2, may be drawn; the lines d, c, a, b being the lines of the piers; g the centre of the under, and h that of the upper arches. The various parts of an arch are shown in

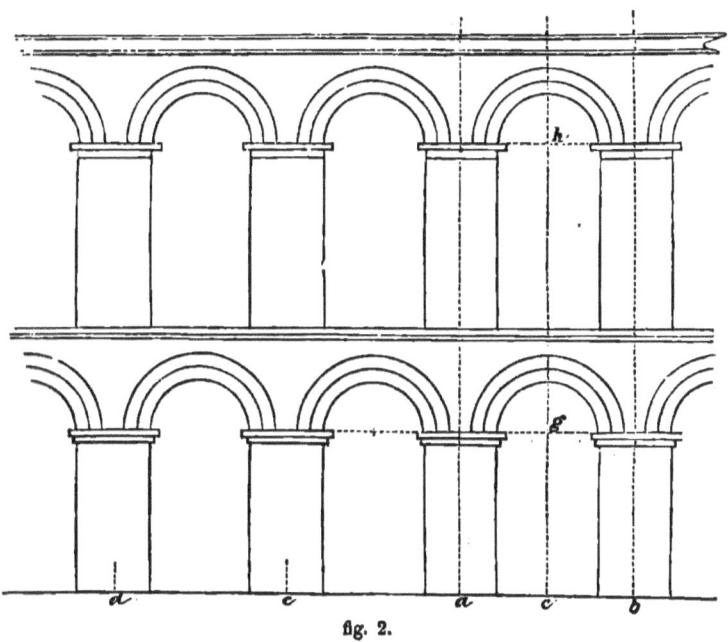

fig. 2.

EXAMPLE 3, fig. 3, where $a b$ is the 'span' of the arch; $e d$ its 'rise;'

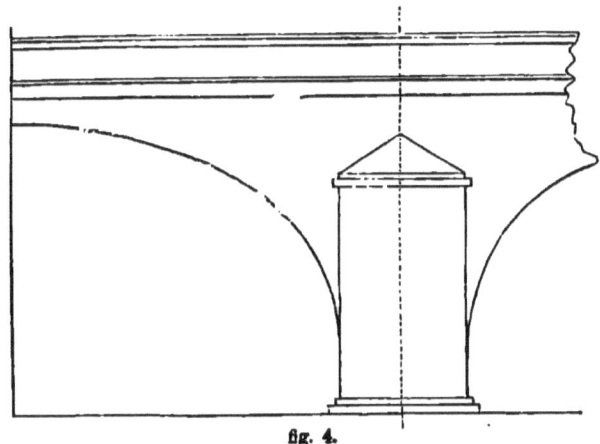

fig. 4.

$a d b$ the inside curve, called the 'soffit,' or 'intrados;' the key-stone is g.

The exterior or upper curve of the arch is called the 'extrados.

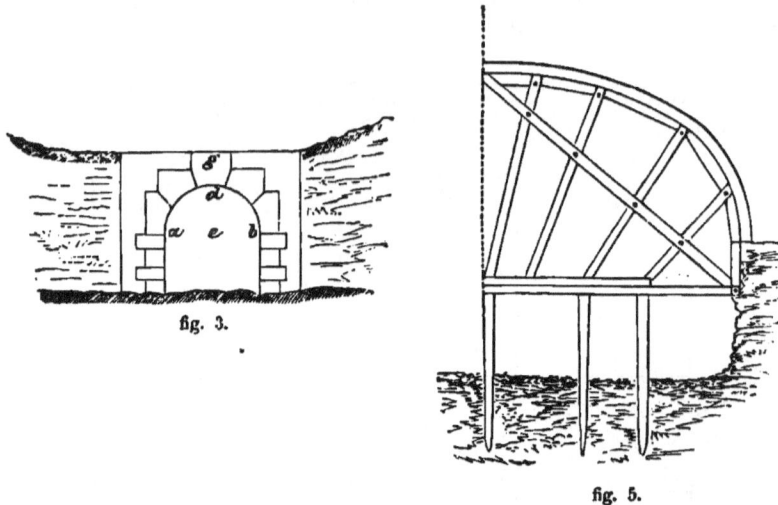

fig. 3.

fig. 5.

EXAMPLE 4, fig. 4, is an elevation of bridge with semi-elliptical arch. For method of describing this form see *Practical Geometry*.

EXAMPLE 5, fig. 5, is elevation of the timber framing or 'centering' of a bridge.

The method of delineating the various features of a country or district in a map is shown in

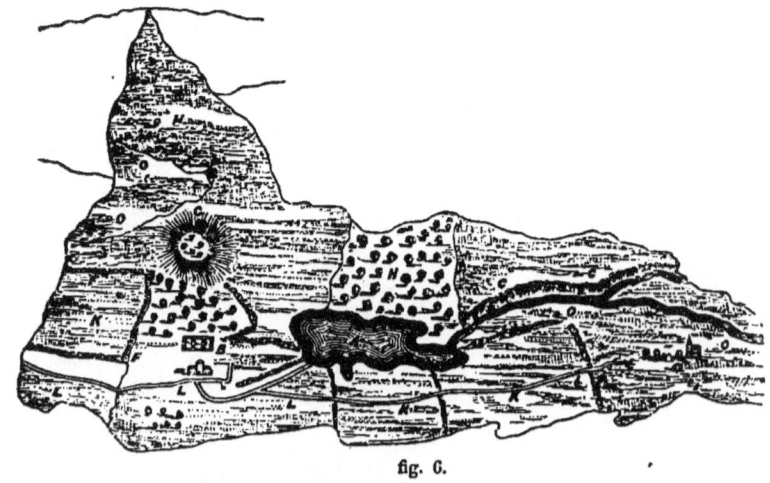

fig. 6.

EXAMPLE 6, fig. 6, where A represents a piece of inland water or lake; E E a river, proceeding from this; B the garden attached to the mansion;

c a hill, with trees on its summit; c c, near the river E E, represents rising ground on its margin; H H plantations of trees; o o a swamp or morass; K K meadow-lands; L L a public highway. In the following illustrations the features are shown on a larger scale, as in

EXAMPLE 7, fig. 7, which represents a hilly or mountainous ridge.

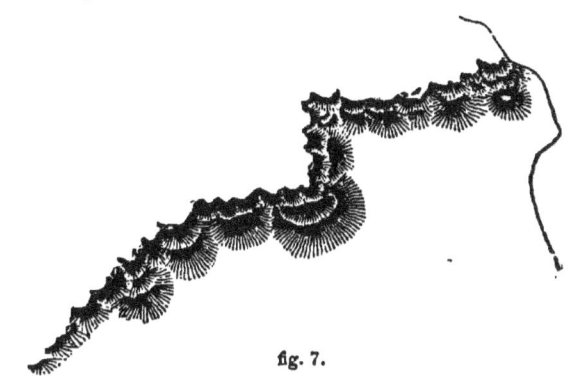

fig. 7.

EXAMPLE 8, fig. 8. Rising ground near a river.
EXAMPLE 9, fig. 9. The same.

fig 8.   fig. 9.

EXAMPLE 10, fig. 10, represents a river, with small stream issuing from it and traversing a meadow. In copying this, the pupil should fill up the whole of the part representing the extent of meadow (within the boundary-line) as in the corner of the illustration now given.

EXAMPLE 11, fig. 11, represents swampy ground with trees.

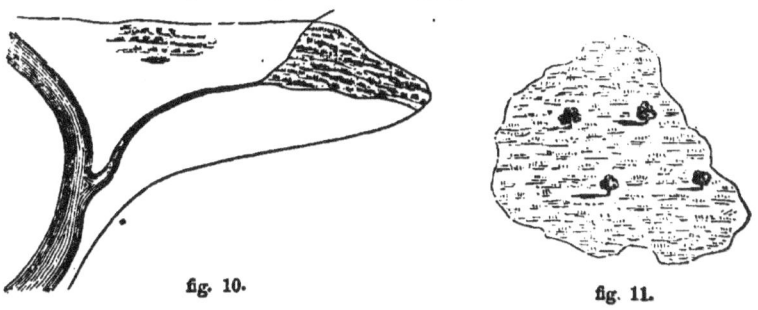

fig. 10.   fig. 11.

EXAMPLE 12, fig. 12, represents a river entering the sea; the coast is delineated as in the sketch.

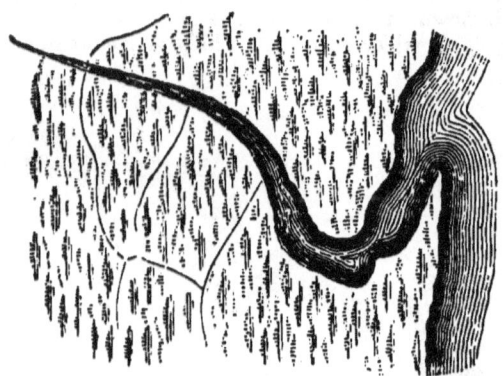

fig. 12.

EXAMPLE 13, fig. 13, represents part of a sea-line of coast cc, with sandy shoal bb, and swampy morass aa.

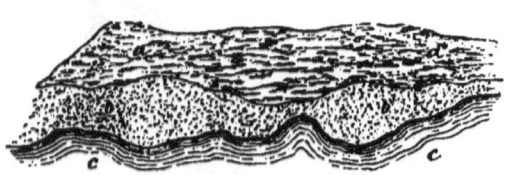 

fig. 13. fig. 14.

EXAMPLE 14, fig. 14, represents the method of delineating a rock, used in marine maps. A range of rocks is represented in
EXAMPLE 15, fig. 15, and a rock surrounded by sand in
EXAMPLE 16, fig. 16.

 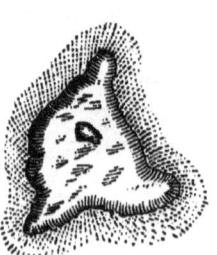 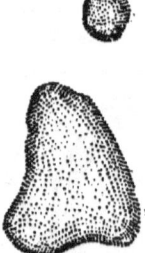

fig. 15. fig. 16. fig. 17.

EXAMPLE 17, fig. 17, represents a sandy shoal. The method of delineating water in a basin or harbour is shown in

110   ILLUSTRATED ARCHITECTURAL,

EXAMPLE 18, fig. 18. The manner of representing blocks of houses in a town or suburban district map is represented in

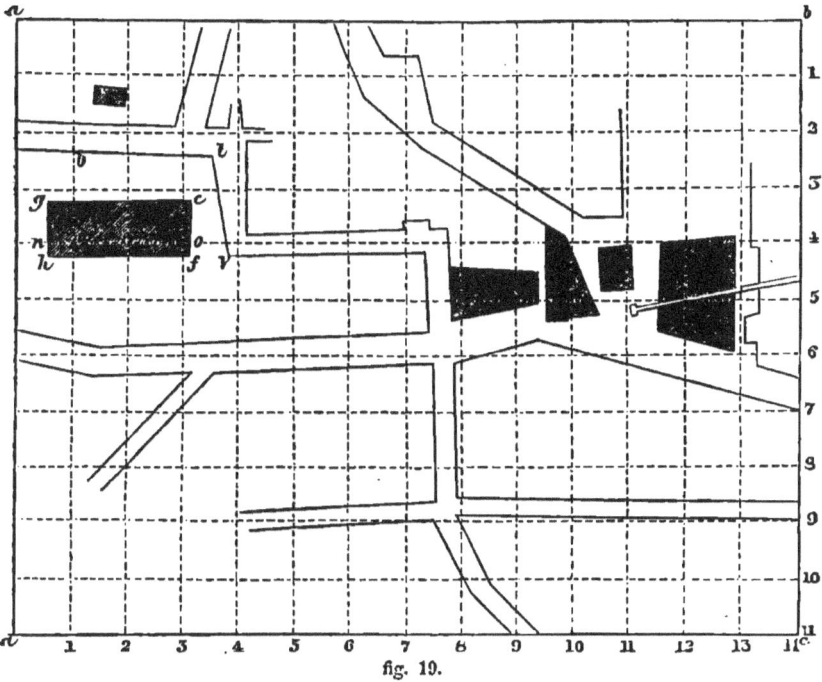

fig. 19.

EXAMPLE 19, fig. 19. This example is also designed to show the use of squares in reducing or enlarging maps. The principle of this method has been fully described in *Practical Geometry*.

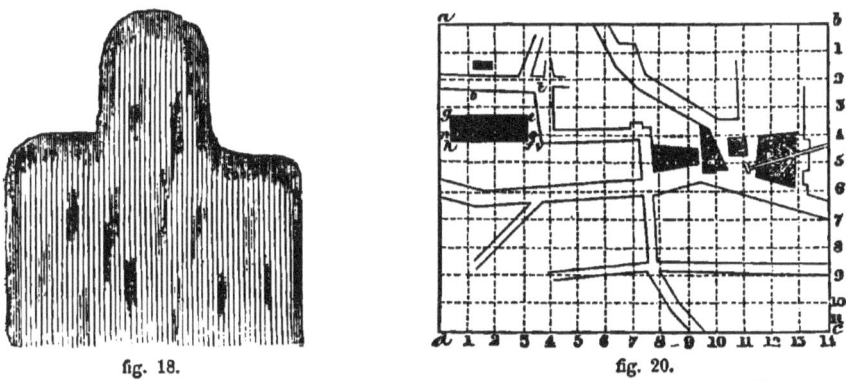

fig. 18.   fig. 20.

EXAMPLE 20, fig. 20, is the same subject as in the previous figure. The pupil, aided by the letters of reference and the figures, should have no difficulty in finding the various points in fig. 20 from fig. 19, and *vice versâ*, if the plan is fig. 20 to be enlarged twice, as in fig. 19. Irregular portions of maps may be copied by adopting offset lines, as in

EXAMPLE 21, fig. 21, which represents part of a river, which is required to be copied and enlarged as below. Draw any line cd; from any scale set off distances, as $cg = 50$, $gh = 60$, and so on. Next draw a line, as po, corresponding to cd; from p set off distances corresponding to those in cd, but taken from a scale larger than that of cd. From the same scale as that from which the measurements on cd were taken, measure the lines drawn at the various points at right angles to cd to where they touch the outline of the lowest side of river, as $g = 40$. Make the line t the same distance, but taken from its proper scale; by proceeding thus, points will be found, by tracing through which, an outline will be obtained equal to that of the copy. The angle dcb is equal to 40°. The pupil should extend this principle of copying irregular figures, by which he will be enabled to judge of its utility in practice.

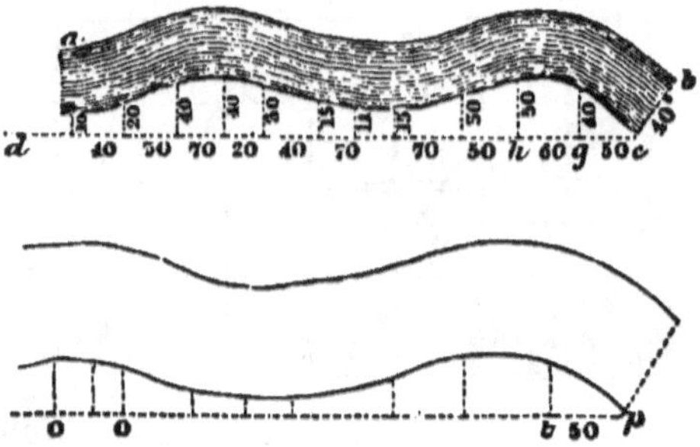

We now give a few examples of the lettering attached to maps and plans.

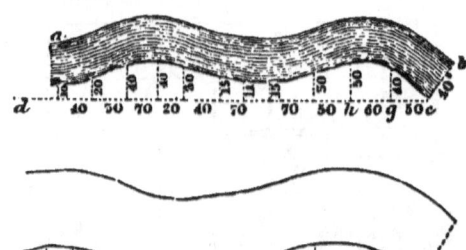

fig. 21.

fig. 22.

fig. 23.

REFERENCES.

GREEN..........

RED.............. fig. 24.

PARISH OF

fig. 25.

EXAMPLE 22, fig. 26, shows the compass-mark in plans, by which the directions are obtained. The *fleur-de-lis* always points to the north.

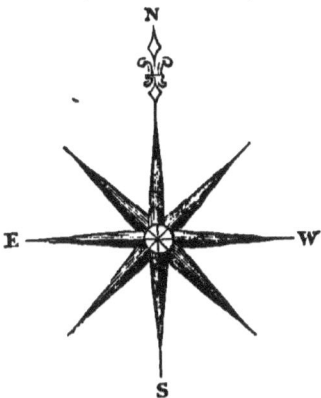

EXAMPLE 23, fig. 27, represents the plan of part of a district through which a road *a b* is to be cut. The section of this is in

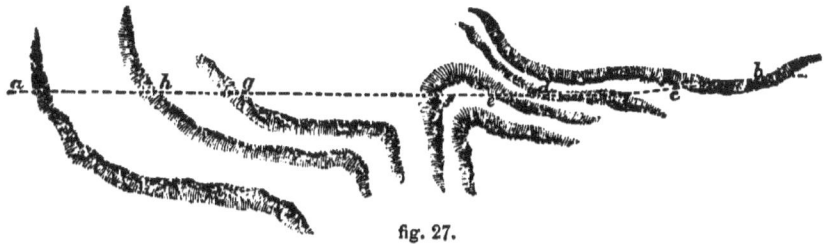

fig. 27.

EXAMPLE 24, fig. 28. The parts filled in with small dots represent hollows filled up; the cross-lines point where a cutting is made. The horizontal line *c d* is termed the 'datum line.' See article 'Levelling' in the work on *Practical Mathematics* in this series.

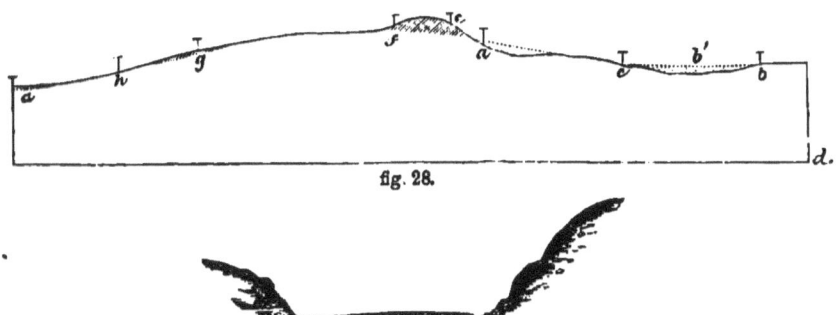

fig. 28.

fig. 29.

EXAMPLE 25, fig. 29, represents a section of road, showing method of delineating it.

EXAMPLE 26, fig. 30, represents the rocks at the side of a section of a railway cutting.

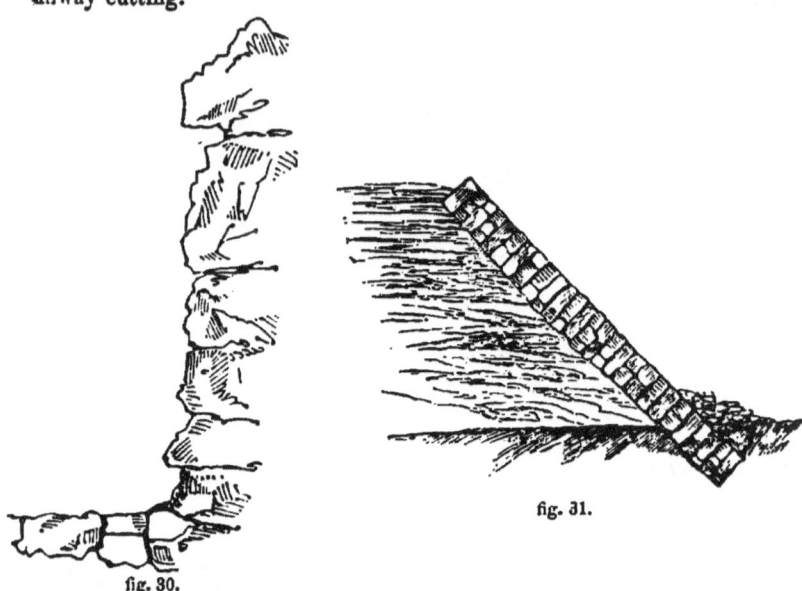

fig. 31.

fig. 30.

EXAMPLE 27, fig. 31, represents the method of delineating an embankment faced with rubble masonry.

EXAMPLE 28, fig. 32, represents a breakwater formed of large stones thrown together, sloping outwards to resist the action of the waves.

EXAMPLE 29, fig. 33, is the section of a stone pier, where aa is the face toward harbour; bb that to the sea; the interior is filled with round stones, as cc. The plan of a retaining wall is shown in

fig. 32.

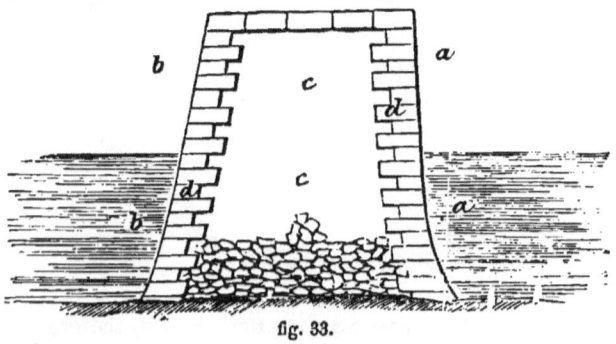

fig. 33.

EXAMPLE 30, fig. 34, where *b c c* is the stone facing; *d* the stones used for filling up.

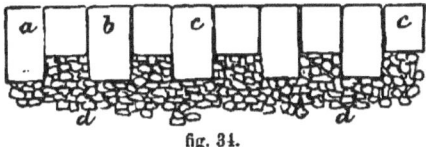

fig. 34.

EXAMPLE 31, fig. 35, represents the footings B of a pier of a bridge resting on a sand foundation at A.

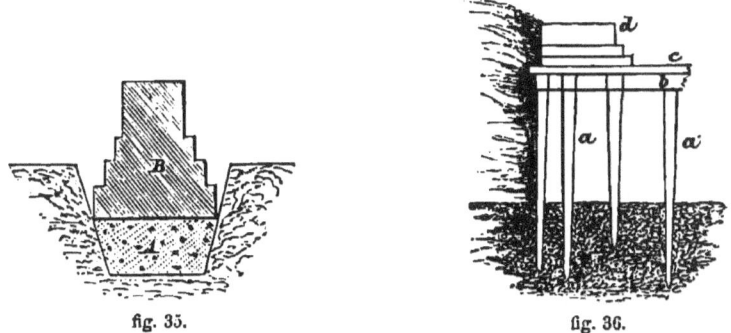

fig. 35.   fig. 36.

EXAMPLE 32, fig. 36, represents piles of wood driven into the ground, supporting masonry. A section of a coffer-dam in a bed of 'beton' is shown in

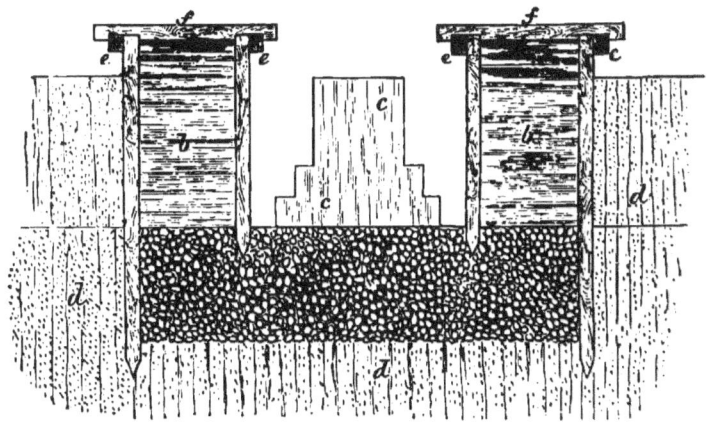

fig. 37.

EXAMPLE 33, fig. 37, where *c c* is the mass of masonry, resting on the mass of beton; *d d* represents mud; *e e* the main piles and 'wales,' and *f f* the cross-pieces; *b* represents the clay-puddling between the piles, which serves to keep out the water from the interior. For explanation

ENGINEERING, AND MECHANICAL DRAWING-BOOK. 115

of the various terms here used, see treatise on *Mechanical and Civil Engineering*.

EXAMPLE 34, fig. 38, is elevation of factory-chimney, of which the transverse vertical section is given in

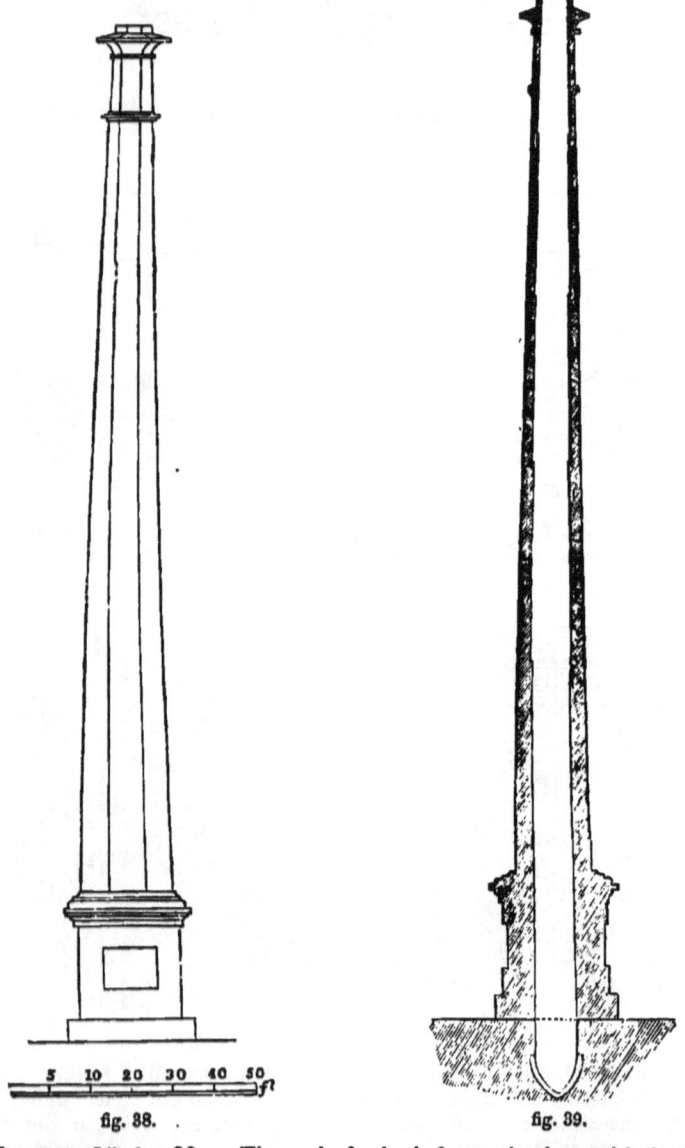

fig. 38.   fig. 39.

EXAMPLE 35, fig. 39.   The scale for both figures is given with fig. 38.

SECTION III.

MECHANICAL DRAWING.

In this department we purpose explaining, by the help of appropriate diagrams, the easiest methods of delineating various portions of machinery. In this, as in the others just treated of, a knowledge of the constructions which we have given in *Practical Geometry* will be essential. The preliminary lessons also of the department of this work on Architectural Drawing will be of use in enabling the pupil readily to master the lessons we now place before him.

EXAMPLE 1, fig. 1. represents a 'bolt,' cb, with the solid head $e'd$, and movable 'nut,' $g'g$. This is used for strongly fastening various portions of machinery together. For examples of the method of using this, see our work on *Mechanics and Mechanism* in this Series. To draw the figure now given:—Suppose the copy to be without the centre-line; bisect $e'e'$ in the point a, draw ab'. On the paper on the drawing-board draw two lines $e'e'$, ab' at right angles to each other; with ae' from the copy measure from the point of intersection of the above lines on the board a to $e'e'$; from a measure to b; from b with distance ae' measure to dd; join de', de'. From a measure to c and b'; from these points with ae' measure to $g'g$, $g'g$; join $g'g'$, gg. From b measure to hh; parallel to ab' from h, h draw lines meeting $g'g$.

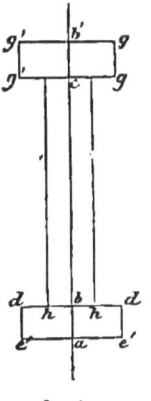

fig. 1.

EXAMPLE 2, fig. 2. Bisect the line $b'b'$ of the copy in the point a', and draw $a'b$. On the paper on the board draw two lines corresponding to these, intersecting at the point a'.* From a' measure to b', b', from a' measure to c; with $a'b'$ from this point measure to ff; draw a line parallel to $b'b'$ through e; join fb', fb'. From a or c measure to d, and through this draw a line parallel to $b'b'$. From e measure to g, g; join $g'g'$ by perpen-

---

* To avoid repetition, the pupil is requested to observe that, in all the lessons, the centre-lines drawn on the various diagrams must be drawn on the paper on the board, it being understood that where a copy is presented him in this book, or elsewhere, without centre-lines being given on it, that these should be adopted and drawn in faint lines, so that data may be obtained from which to take measurements. By dint of practice the facility for copying without these will be attained, or, at least, they will be sparingly required. As the pupil proceeds, he will the more readily decide as to the quickest method of finding datum points from which to take measurements.

dicular lines to gg on the line ff. From a' measure to e; draw a line through this parallel to $a\,b'$; from e measure to $e'\,e$; from d measure to $h\,h$ on the line $g'\,g'$; join $h\,e'$, $h\,e'$. Where we use the terms 'measure from'—as measure from a' to b—we mean, in all instances, that the measurement $a'\,b$ is to be taken from the copy and transferred to the paper on the board, from the point thereon corresponding to the point a' in the copy. Again, when we say 'measure from a' to b',' we wish the pupil to take the measurement $a'\,b'$ from the copy, transferring it to the line on the paper

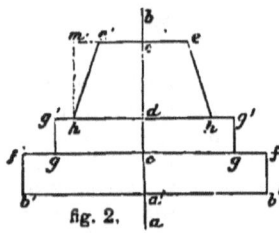

fig. 2.

corresponding to the line $b'\,a\,b'$ in the copy, from the point on the paper corresponding to the point a' in the copy. Hence the pupil will observe the use of datum-lines—as $a\,b$, $b'\,a'\,b'$—from which to take the measurements from the copy; these to be transferred to the paper on the board on which the fac-simile is to be constructed. As a means of enabling the pupil readily to decide on datum-points from which to take measurements, we explain another method of copying the last figure. Draw any line $b\,a'\,b'$, assume any point on it, and draw there a line at right angles to $b'\,a'\,b'$. The intersection of these lines will represent the point b' in the diagram just given. From b' measure to f in the copy, and transfer it from b' to the line which is at right angles to $b'\,a'\,b'$ as to f; from f draw a line parallel to $b'\,a'\,b'$. From b' measure to b', or from f to f'; join bf'. The part $b'f, f''b'$ will thus be put in; the part up to $g'\,d\,g'$ may thus be put in without the use of a centre-line. The part to e can be quickest put in by using one; however, it may be done as follows:—Measure from d to h; from h draw a line to m, at right angles to $g'\,d\,g'$; with $d\,e$ or $a\,e'$ measure to e, and draw through this a line $e'\,e$ parallel to $a'\,b'$. From m measure to e', and from e' to e; join $h\,e'$, $h\,e$. In the following diagram the use of the circle is shown.

EXAMPLE 3, fig. 3. Draw any two lines on the board corresponding to $a\,e$, $g'\,g\,g'$ in the copy. From g measure to b, c, and d; from g measure to

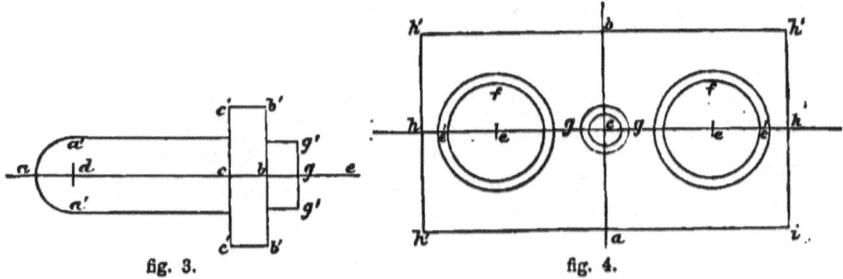

fig. 3. fig. 4.

$g'\,g'$, and from b to $b'\,b'$; join $g'\,g'$ to $b'\,b'$ by lines at right angles to $g'\,g'$. From c measure to $c'\,c'$; join $b'\,c'$, $b'\,c'$. From d, with $d\,a'$ as radius, describe a semicircle $d\,a'\,a'$; by lines parallel to $c\,b$ join $a'\,a'$ with the line $c'\,c'$.

EXAMPLE 4, fig. 4. Draw on the board two lines corresponding to $a\,b$, $h\,h$ in the copy. From the point of intersection c measure to $a\,b$, and $h\,h$; through $a\,b$ parallel to $h\,h$ draw lines $k'\,k'$, $k'\,i$; through $h\,h$ parallel to $a\,b$

draw lines meeting those in the points $h'\, h'$, $h'\, i$. From c with cg put in the circle; from c measure to e, e. From these points, with $e'\, e'$ as radius, describe the circles, and also the interior ones, as ef.

EXAMPLE 5, fig. 5. Draw on the board, lines ab, cc at right angles, intersecting at e, corresponding to those in the copy. From e measure to a

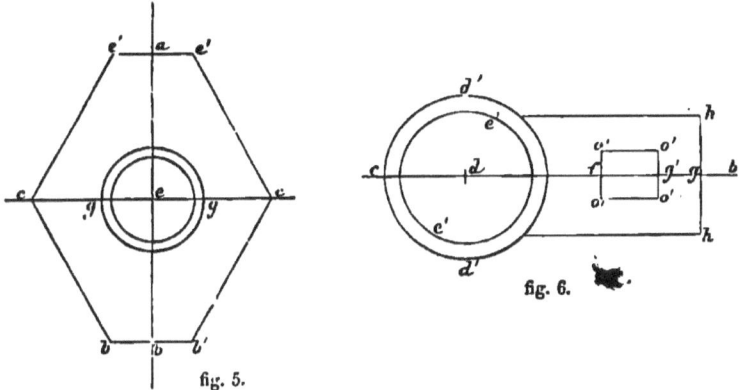

fig. 5.

fig. 6.

and b; from these points draw lines parallel to cc, from ab measure to $b\, b'$, $c'\, c'$. From e measure to cc; join ce', cb', ce', cb'. The radius of the circle in the centre is eg.

EXAMPLE 6, fig. 6. Draw lines corresponding to $b\, d\, c$, $h\, h$ in the copy. From g measure to d; put in from d as a centre, the circles $d'\, d'$ and $e'\, e'$. From g measure to h, h, and parallel to bd from these draw lines touching the circle, $d'\, d'$. From g measure to g' and f; from these points measure to $o'\, o'$; through $o'\, o'$ draw lines parallel to hh and to bd.

EXAMPLE 7, fig. 7, represents a set of what are termed 'speed pulleys' (see *Mechanics and Mechanism*). Draw any two lines corresponding to ab,

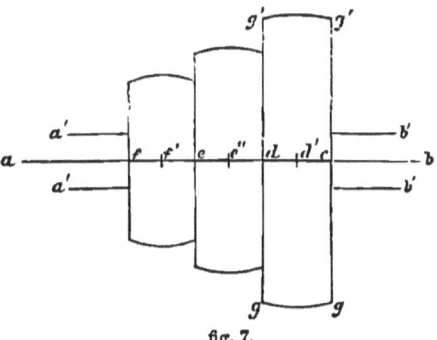

fig. 7.

$g'\, g$. From c measure to d, through this draw a line parallel to $g\, g'$; measure from c and d to g', g. Bisect the distance $d\, c$ in d'; from d' as a centre, with $d'\, g'$ as radius, describe the arcs joining the lines through $g\, g'$, $g'\, g$. In like manner, measure from c to e and f: the points f, e' will be the centres of the arcs joining the lines drawn through e and f.

Example 8, fig. 8, represents a projecting 'snug,' by which two parts may be joined by means of a bolt secured by a nut, passed through holes bored in each. Draw the line ab, and another at right angles to it. From a measure to b, and put in the various horizontal lines and the base; from b measure to c, and parallel to ab draw a line from this point. From c measure to d; from d as centre with radius dc describe the curve. From f measure to e; a line drawn from this, parallel to ab, gives the end-line. The centre g (as also d) is found by trial on the copy, and the points transferred to corresponding parts on the board. The line dc represents one method of transferring them.

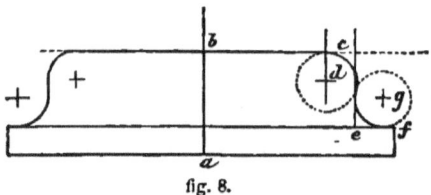

fig. 8.

Example 9, fig. 9, represents a side view of a 'pulley,' or 'drum,' showing the arms and centre. Draw any two lines corresponding to ab,

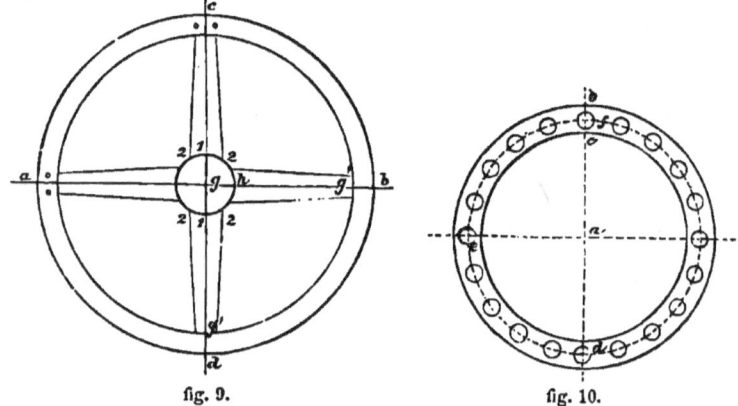

fig. 9. fig. 10.

cd. From g as centre, with gb as radius, describe the circle, and also the interior circle $g'g'$; from g with gh put in the small circle representing the diameter of the centre or eye of the wheel. From the lines 1, 1 with distance 1, 2 lay off on either side of all the centre-lines of the arms; next, from the points where the interior circle cuts these lines at the points g', g', lay off on each side equal to half the thickness of the end of the arm as it joins the inside of wheel. Join the points thus obtained with those previously obtained on the centre of the wheel, as 2, 2.

Example 10, fig. 10, represents the plan of a circular cylinder or receptacle, the small circles showing the position of the circular heads of the bolts used for attaching the cover to the main body of the receptacle. The method of finding the centres of the small circles is as follows: Draw any two lines ae, bd; from the point of intersection as centre, with radius ab, ac, describe circles; bisect the distance between these, as bc, in the point f. From a as centre, with af as radius, describe a circle fed: the centres of the small circles will be found on this line. Find the position of any two of the circles, as fe or ed; transfer these points to the board. In the copy, the centres of four of the circles will be found where the diameters ea, bd cut the circle drawn through fd. Count the number of circles between f and e, or e and d; divide the circular line passing through f, and between e and f or e and d, into as many equal parts as will give as many centres as there are circles in the copy: these points will be the centres of the circles.

EXAMPLE 11, fig. 11, represents the plan of a small thumb-wheel attached to the head of a screw-bolt, by which it may be easily moved by means of the finger and thumb. From a with ab describe a circle, draw the diameter db; divide the semicircle db into four equal parts in the points ef; from a draw lines through ef; and continue these to cut the other semicircle. From a measure to n, the centre of the circles forming the ends. With an describe a circle: the points on the radial lines, as n,

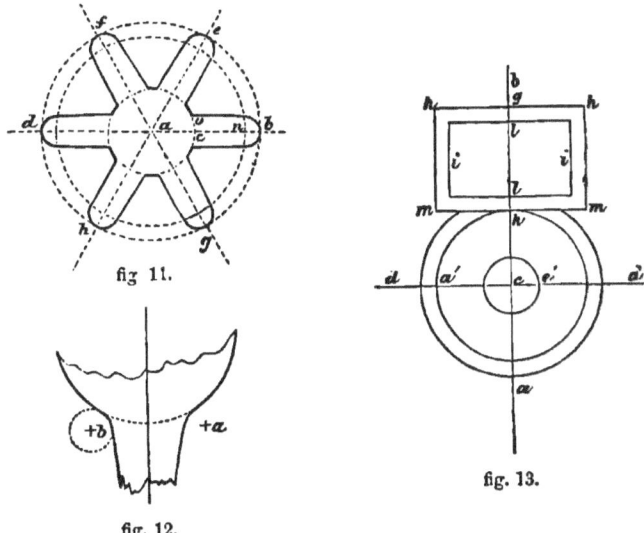

fig. 11.

fig. 12.

fig. 13.

where this intersects them, are the centres of the circles which terminate each radial arm. From a describe the small circle ac; from the points where this intersects the radial lines, as c, lay off on each side of these the distance co; join the points thus obtained on the circle aco with the extremities of the circular ends. Another way of joining the radial arms to the centre or eye may be understood by inspection of the diagram in fig. 12, where ab are the centres of the circles, part of which joins the arm with the centre.

EXAMPLE 12, fig. 13. Draw any two lines corresponding to ag, dd in the copy; from the point of intersection c measure to the points h, g; through these draw lines parallel to dd. From h, g measure to mm, hh; join mh; put in, in like manner, the internal parallelogram $lili$. From the point c, with radius ce', ca', and ca, describe the circles as in the copy, meeting the line mm.

ENGINEERING, AND MECHANICAL DRAWING-BOOK.   121

EXAMPLE 13, fig. 14, represents plan of part of a 'valve-plate.' From any centre *a* describe a circle *a b*, and one within this, as *a c*; continue this last all round, the part from *m* to *p* being afterwards rubbed out when the drawing is finished and inked in.

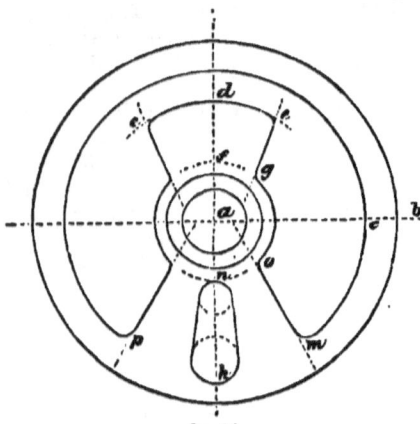

fig. 14.

From *a* with *a d* put in part of a circle *e d e*. From *d* measure to *e, e*, and through these draw lines to the points, as *g*, on each side of the line *f*. On each side of the line *a h* measure to *p* and *m*, also from *n* to *o*; join *m o*. Put in the circles at *n* and *h*; join them as in the drawing.

EXAMPLE 14, fig. 15, represents the plan of a 'lever.' Describe the circle *a h*, draw through *a* the diameter *b a d*; from *a* measure to *c*; put in the circle *c d*. Bisect *a c* in *e*, and through this draw a line at right angles to *a d*, as *ff*. In the copy take the points *f* (where *ef* intersects the curve), *h* and *g* (where the curve *h g* touches or joins to the circles described from *c* and *d*). By means of these points (see the problem in the work on *Practical Geometry*, to find the centre of a curve, three points in that curve being given), the centre *m* will be found.

EXAMPLE 15, fig. 16, represents the method generally employed of constructing the central part of a "spur-wheel." The circles *c, f*, and *m* are

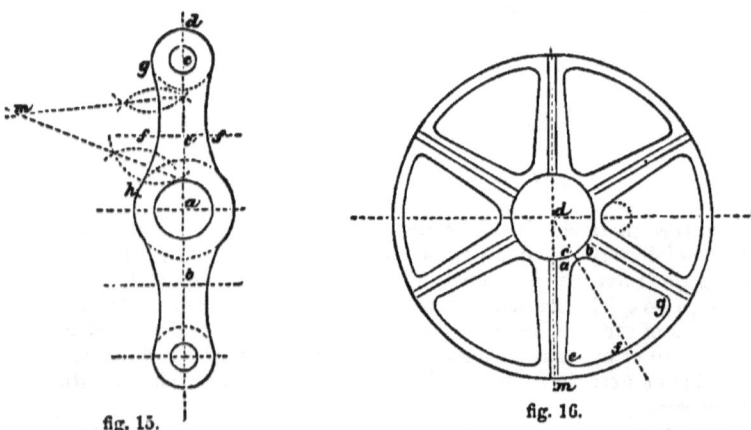

fig. 15.   fig. 16.

described from the centre *d*; the circle *m* is divided into as many equal parts as there are arms in the wheel, any central point of these, as *m*, being adopted as the datum-point from which to take the measurements. The space between any two of these arms, as *a b*, is bisected, and a line, as df, drawn. By measuring from f to e, g, the centres of the curves at e and g will be obtained; the centre of the curve $a\,b$ is also on the line df.

EXAMPLE 16, fig. 17, represents the plan of a pulley with curved arms. The method of describing these is explained in

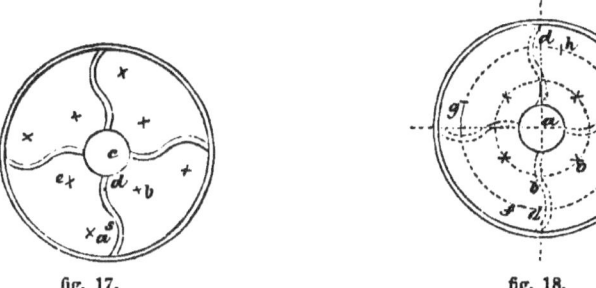

fig. 17.   fig. 18.

EXAMPLE 17, fig. 18. The first operation necessary to be done is to find in the copy fig. 17, the centres of the circles forming the curves: these must be found by trial. Next draw two lines at right angles, as in fig. 18, intersecting in the point a corresponding with the centre c, fig. 17. From a describe circles representing the rim and the eye of the wheel in last figure. From c, in fig. 17, measure to the centre b, from which the curve d is described, and from a, fig. 18, a circle $a\,o$: on this line the other centres, as e, fig. 17, will be found. In like manner, from the centre c, fig. 17, measure to a, from which the curve $a\,s$ is described, and from a, fig. 18, describe the circle $g\,h$. On this will be found the second set of centres. From d measure to h, from h to n, from n to f, and from f to g: these are the various centres. Or the curves next the eye may be drawn in first, and the curves with radius $a\,s$ be described, to meet these from the circle $g\,h$. In this example the arms are of uniform breadth; where they get gradually less from the centre or eye of the pulley outwards, the method of describing them may be learned from

EXAMPLE 18, fig. 19. The points from which the curves are drawn must be found, and corresponding points transferred to the paper, as in last example. Two circles, as d, o, will thus be obtained, in which the centres of the various curves will be found. Put in the circle representing the eye of the pulley, and draw a diameter $a\,b$; draw a line in the copy corresponding to this, and measure from b to the point representing the centre of the circle from which the curve $c\,c$ is drawn, as d; transfer this to the copy, and from d with $d\,c$ draw the curve $c\,c$; from c measure to f, thus giving the breadth of arm at eye; from f, with the radius of the curve f taken from the copy, cut the circle o in o; from this point with same radius describe the curve fg. The various points denoting the centres of the curves are given in the circles, the points $e\,e$ being those where the curves join the central circle or eye of the pulley.

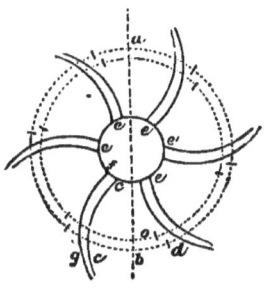

fig. 19.

EXAMPLE 19, fig. 20, represents the bottom part of foot of a cast-iron framing. Draw a line cd; from c measure to a and b; through these

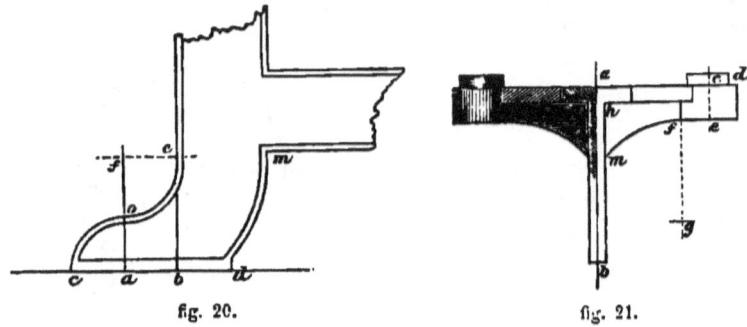

fig. 20.   fig. 21.

draw lines perpendicular to cd; with ac from a describe the curve co. From b measure to e. Find the centre of the curve joining oe, at f. Find by any of the methods already described the point m; join md by the curve.

EXAMPLE 20, fig. 21, represents part of the frame-work forming the support for the bearings c in which vertical spindles revolve. Draw ab, ad; measure from a to d and c; draw ce at right angles to ad. From e measure to f; and from f draw to g parallel to ab; from a measure to h and m. The centre of the curve joining fm will be found at g on the line fg. The method of filling-in the drawing is shown by the other half.

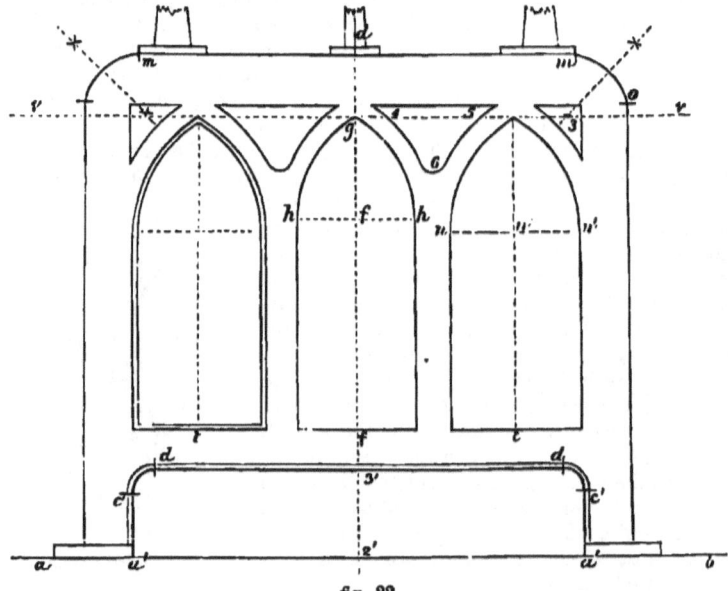

fig. 22.

EXAMPLE 21, fig. 22, represents the outline of side elevation of framing. Draw the line ab, and at right angles to it $2'd$; measure from $2'$ to $a'a'$, and to $3'$. Through these points draw lines dd, $a'c'$, $a'c'$; join the points c', d by the part of the circle, as in the diagram. From $2'$ measure to f, and draw the line tft; from f measure to t, t; from these points draw lines parallel to $2'd$. From t measure to n; draw nn', and from n, n', with radius nn', describe curves meeting, as in the drawing. From f measure to f, and draw hfh; from h, h with radius hh describe curves meeting in g on the line vv. The curves 5, 6 and 3 are described from the centres n', n, and 4, 6 from centre h on the left hand side of ff. The lines mm, oo are joined by curves described from the centre 3, which centre is found by describing arcs from the points m, o with any radius greater than half mo, and joining the intersection of these arcs by a line as in the copy.

EXAMPLE 22, fig. 23, is another outline representing the side elevation of framing. The curve h is described from the centre f on the centre-line bf; the centre-lines of the other parts are at m, o, d, and c.

EXAMPLE 23, fig. 24, is another form of framing. The centre of the curve n, joining the lines from m, m, is at h, on the centre-line oh; the centres d, d are on the line drawn through c to hb, parallel to mm; the centre of the circle c is at g.

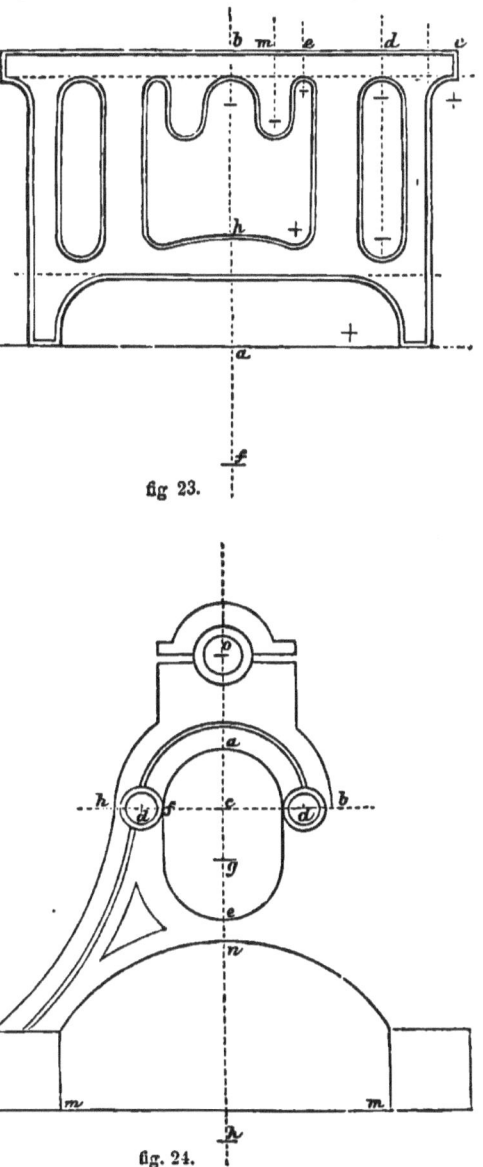

fig. 23.

fig. 24.

EXAMPLE 24, fig. 25, represents the front elevation of a 'cross head' and 'side levers.' The centre-lines are $a\,d$, $e\,h$, $v\,v$. The plan is shown

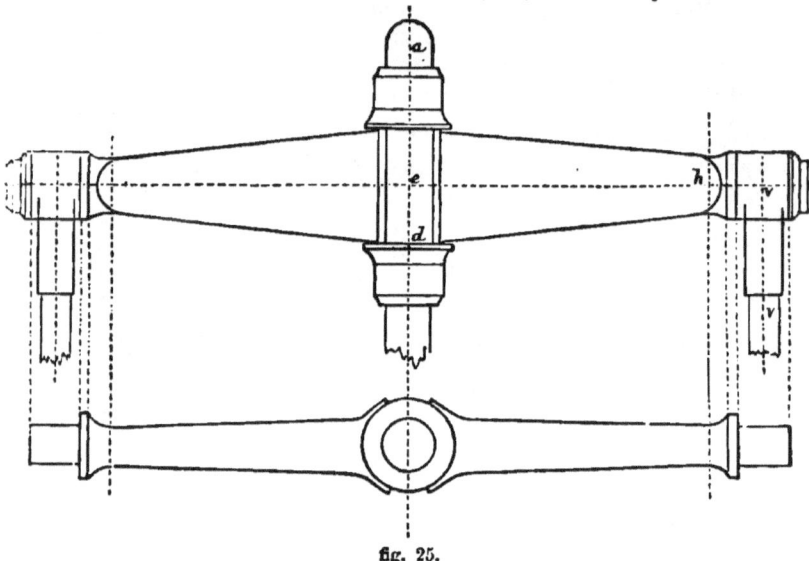

fig. 25.

below, the lines of which are obtained by continuing those of the upper figure, as in the drawing.

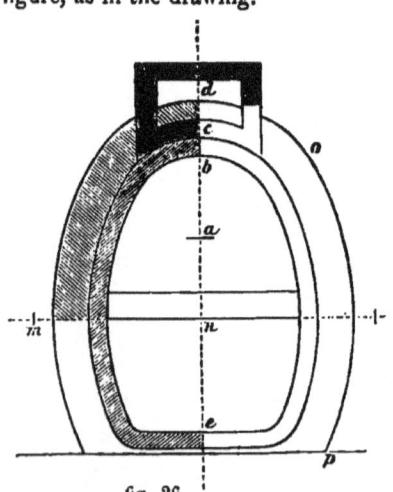

fig. 26.

EXAMPLE 25, fig. 26, represents the front elevation of the cover for a gas retort. The centre of the parts b, c, and d is at a on the line $d\,e$; the centre of the curve joining $o\,p$ at m, on the line $n\,m$.

EXAMPLE 26, fig. 27, represents the 'transverse vertical section' of a boiler $a\,b$, and its brick 'setting.' From a with $a\,b$ describe the circle $a\,b$; from a measure to c: draw $c\,d$, and from d, $d\,e$. From d measure to g, from which point a line drawn parallel to $c\,d$ marks the point f, where the curve $f\,o$ terminates at the boiler. The point n' is the centre of the curve $f\,o'$; transfer this part from f to n', and describe $o'\,f$. From a measure to the lines $o\,s$, $n\,m$, and draw lines through these parallel to $c\,d$; measure from d to r and g. The centre of the curve $o'\,h$ is at s, and that of the curve $h\,r$ at m.

EXAMPLE 27, fig. 28, represents an 'angular-threaded screw.' To copy it, proceed as follows: Measure from a to d, and from d to e, 1, 2, 3, &c.

These are the points through which the centre-lines of each thread are drawn. From a measure to f, and draw fg; and from a to b and c, and

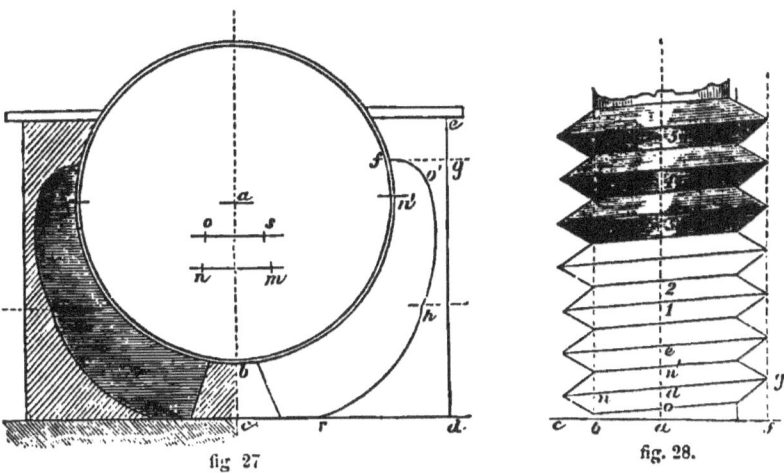

fig. 27. fig. 28.

draw bn. From f on the line bf measure to g, and from b to n; through d draw $n\,d\,g$, and parallel to this, through e, 1, 2, 3, &c., draw lines. Next, from d measure on each side of $d\,g$, equal to half the breadth of each thread, to n'. These lines terminate at the perpendicular bn· join the angles as in the drawing.

EXAMPLE 28, fig. 29, represents a 'square-threaded screw.' From e measure to a; ab, bc, cd, represent the thickness of each thread and the

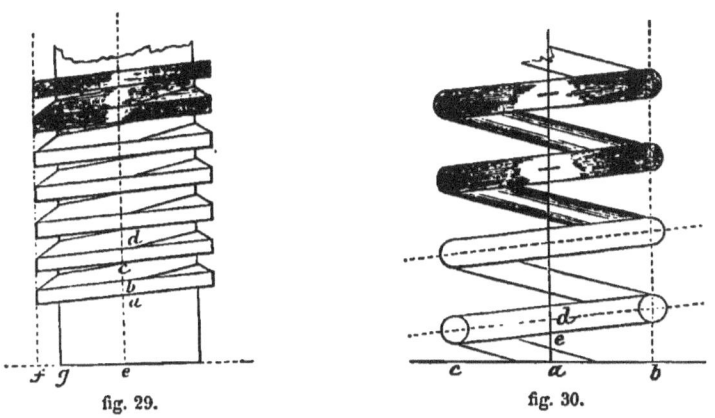

fig. 29. fig. 30.

distance between them; the line from g is the line of the inside of the screw, the line f the outside line of the threads. The last example shows the method of copying this.

EXAMPLE 29, fig. 30, represents a 'helix' of wire, ad being centre-line, de being half the thickness of the coil, the lines from c, b intersecting those drawn parallel to *d*, giving the centre of the circles forming the termination of coils.

EXAMPLE 30, fig. 31, represents another form of screw.

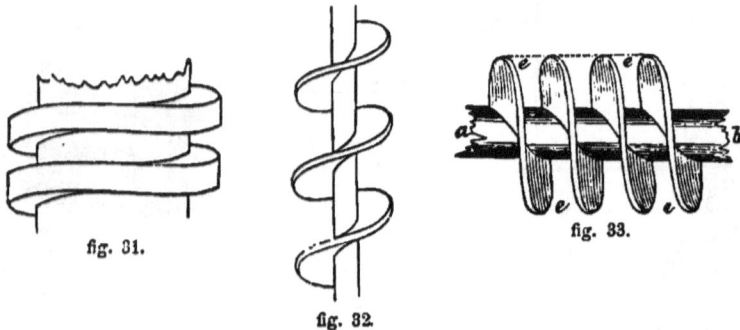

fig. 31.

fig. 32.

fig. 33.

EXAMPLE 31, fig. 32, represents the Archimedean, or endless screw; and another form is given in

EXAMPLE 32, fig. 33, where *a b* is the central shaft round which the helix or thread *c c* is coiled, according to a determined pitch.

EXAMPLE 33, fig. 34, shows the method of drawing-in the teeth of wheels. Let *c x* be the diameter of wheel from centre to outside of teeth.

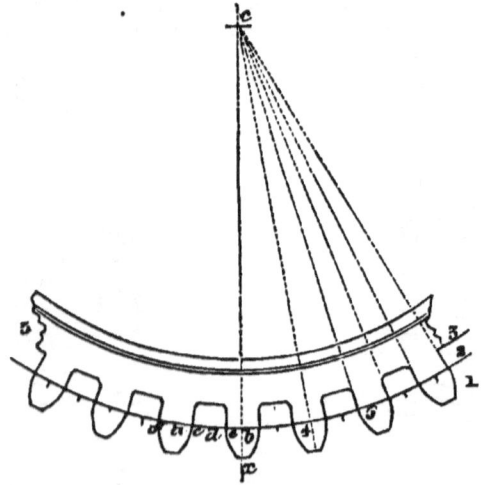

fig. 34.

The circle, of which part is shown, and of which *c b* is the radius, is termed the 'pitch-circle or line.' It is on this line that the number of teeth are marked off. Having ascertained the diameter of pitch-line, the depth of teeth, and the number of them, divide the pitch-circle into as many equal parts as there are to be teeth in the wheel, and proceed as follows : Let *a*, *b*, 4, 5, &c., be the divisions on the pitch-circle representing the centres of teeth; divide the distances between them into two equal parts, as at *d*. From *d* as a centre, with *d b* on both sides of the point *d*, describe arcs of circles as *f b*, joining the pitch-circle and the outer circle, giving the termination of the teeth as the circle *x* 1. Proceed in this way till all the arcs are made to join the circle *x* 1, 2 *d*. The bottom of the teeth are formed by radial lines drawn as from *c e* to the centre *c*, as in the diagram. The forms of teeth are various (see treatise on Mechanics in this series). For the method of describing different curves, and of setting out teeth of wheels and pinions, see treatise on Mechanical and Civil Engineering. The method of drawing the side elevations of toothed wheels may be seen in

EXAMPLE 34, fig. 35. The small dotted circles show another method of describing the form of teeth. The manner of delineating bevil-wheels

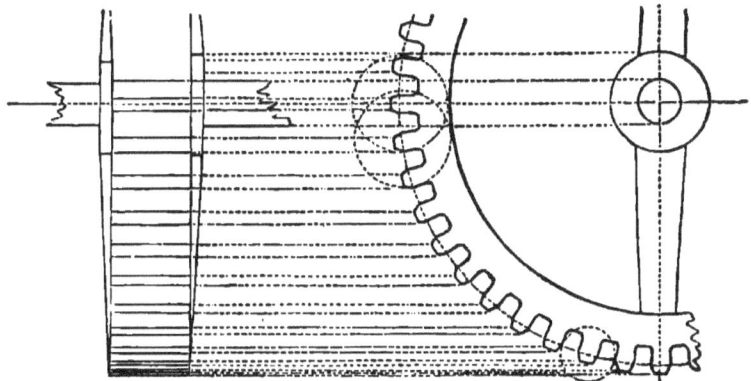

fig. 35.

(for the nature and operation of these see treatise on Mechanics in this series, at pp. 51, 52), may be gathered from the two following figures.

EXAMPLE 35, fig. 36. Let *a b* represent the centre-line of the wheel, *c d* the line of its greatest diameter or 'pitch-line,' *f* the line giving termination of teeth, *d m* being the breadth of the teeth. The teeth on the part between *c v*, *d m* converge to the point *b ;* those between *k d*, *c n* to the point *a*, on the line *a h g*, *e f b*. It is foreign to the purpose of this work to go into the subject of the teeth of wheels, belonging, as it does, to a strictly technical department; we cordially recommend, however, to the pupil anxious to study this interesting and important department, Buchanan's work on *Mills and Mill Gearing*, edited by Sir John Rennie, and published by Weale of Holborn; and the *Engineers' and Machinists' Assistant*, by Blackie of London and Glasgow. Both of these works, although somewhat high-priced, abound in valuable information. We may possibly, at some future time, publish a companion to this treatise, which may serve as a guide or introduction to the sciences of Civil and Mechanical Engineering. To proceed, however, with our explanation. The method of copying the teeth of bevil-wheels may be seen in

EXAMPLE 36, fig. 37, where *a b* is the centre-line of wheel, *c g* the pitchline, *e h* the line terminating the teeth on the back part of the wheel *e g*.

The line *x x* gives the termination of the inside of the teeth, *d f* that of the outside; the lines *g o, g f* are projected towards points on the line *a b*, cor-

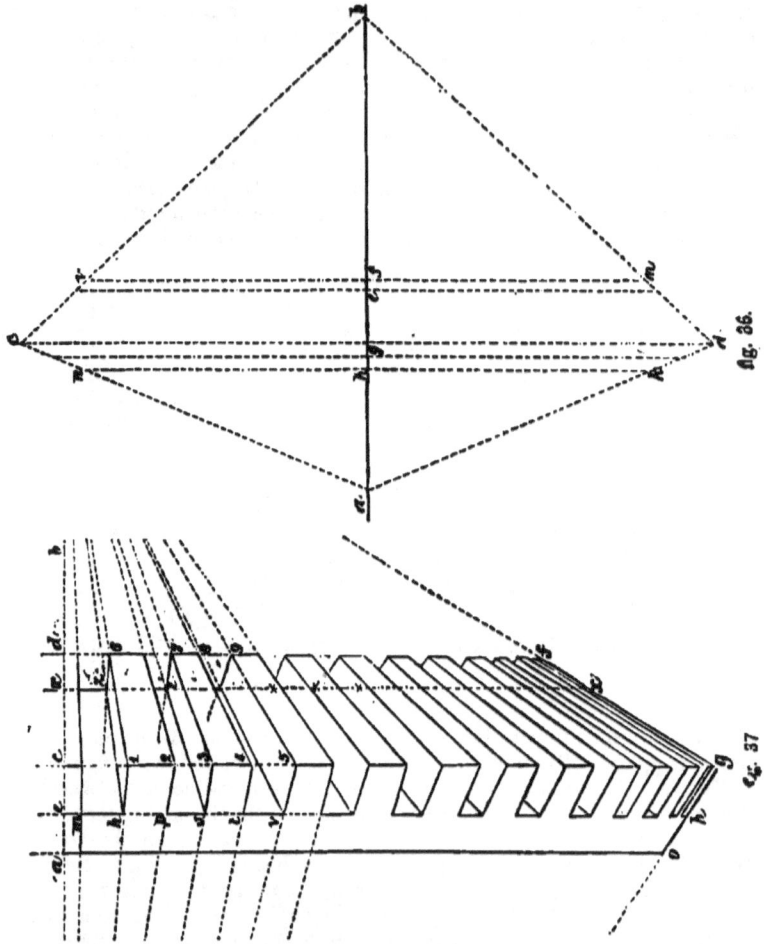

responding to *a l* in fig. 36. The distances between the teeth are set off on the line *e h* to *m, h, p, s, t*, &c ; lines are drawn from these to the point on the line *a b*, to which *o g* converges; these lines are produced to meet the line *c y* in the points 1, 2, 3, 4, 5, &c. From these points, lines, as 1, 6, 3, 7, 5, 9, are drawn to the point on the line *a b*, to which *g f* converges ; these lines are terminated by the line *d f*. From the points *h, s, v,* &c., lines are drawn to the same point on *a b*, as 5, 9, &c., these being terminated by the line *x x* ; the points 6, 7, 9, &c., are then joined to these, as 6 *z*, 2 *t*, &c. The pupil should put in the whole of the wheel, of which only half is here given.

Mechanical drawings are reduced or enlarged quickest by means of what are termed 'proportional compasses.' If these are not available, 'scales' should be drawn from the different figures. Thus, to reduce the drawing in

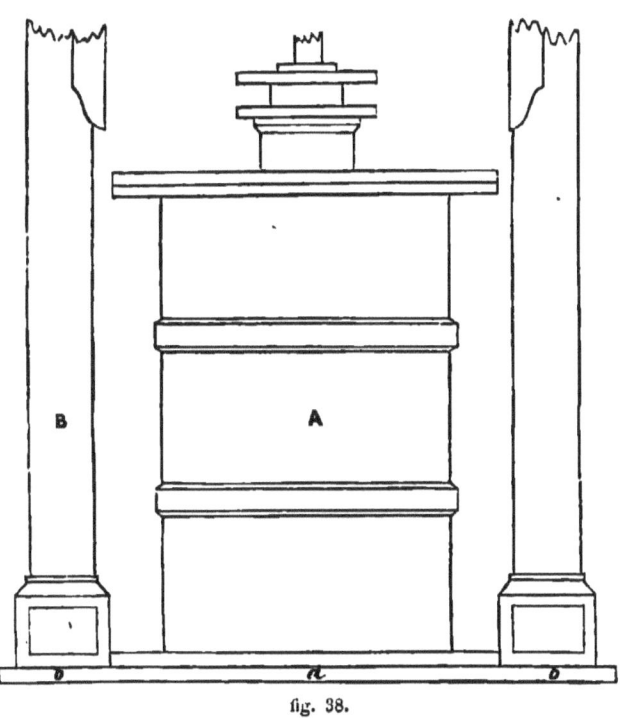

fig. 38.

EXAMPLE 37, fig. 38, of which the scale is given in fig. 39. Suppose the drawing is to be reduced one-half, a scale half fig. 39 is to be made, as in fig. 40; and as each measurement is taken in the compasses from fig. 38, it must be applied to the scale in fig. 39. Suppose this distance is found to be 6 feet, then the distance of 6 feet must be taken from the scale of fig. 40; and the line thus obtained must be drawn in a situation corresponding to that in fig. 38. The result will be a reduced copy, one-half of the size, as shown in

EXAMPLE 38, fig. 41. To reduce by means of the proportional compasses: Having previously set them at the desired mark on the scale attached to each instrument, according to any proportion as desired, all that is necessary to be done is to take any measurement with one end; the distance corresponding to this, reduced or enlarged, is given in the other ends. This being transferred to paper, the desired distance is obtained at once. To reduce by means of the ordinary compasses, without the use of a scale as just described in figs. 38-41, is a matter requiring greater time, and accuracy of adjustment of the compasses is indispensable. Suppose ab, fig. 41, to be the points representing the intersection of the centre-lines

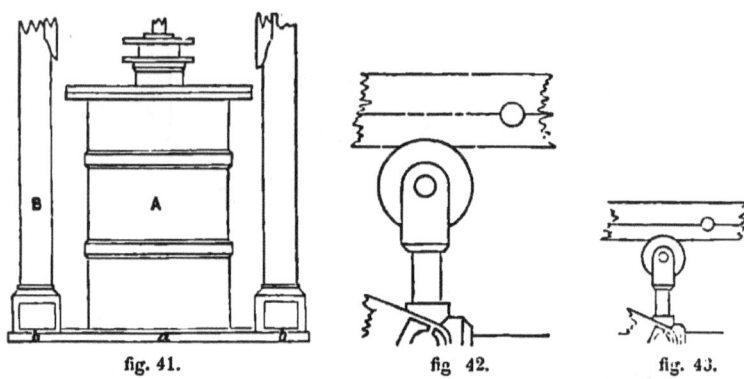

fig. 41.   fig. 42.   fig. 43.

of the parts A, B with the base-line ab, and that a line corresponding to the centre-line from a was drawn on paper, and that half the distance ab in the copy was to be transferred to the paper, half of ab would have to be found in the first place on the copy and transferred. By proceeding thus, a copy of fig. 41, but only half its size, would be obtained. The enlargement of figures is exactly the *converse* of what we have described in figs. 38-41.

EXAMPLE 39, fig. 42, is a drawing which is reduced half in

EXAMPLE 40, fig. 43.

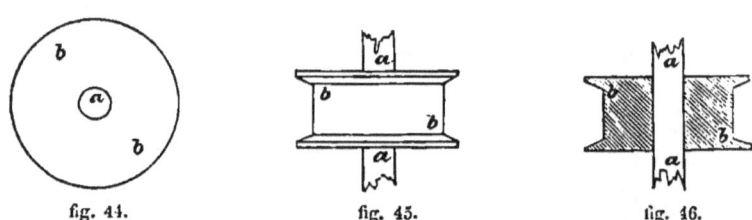

fig. 44.   fig. 45.   fig. 46.

Mechanical drawings are delineated in three ways: as 'plan,' shown in

EXAMPLE 41, fig. 44, which represents the 'plan' of a pulley or solid drum. In 'elevation,' as in

EXAMPLE 42, fig. 45, which is the elevation of fig. 44. Elevations may be 'front,' 'back,' 'end,' or 'side.' In 'section,' as in

EXAMPLE 43, fig. 46, which is a transverse vertical section of figs. 44 and 45. The same letters of reference denote the same parts in these three sketches. Sections may be divided into 'transverse' and 'longitudinal,' these being either vertical or horizontal.

In finished outline-drawings, shadow-lines are made use of. The light, in the generality of examples, is supposed to come from the top and left-hand side of the drawing, thus throwing the right hand and under lines

132   ILLUSTRATED ARCHITECTURAL,

in shadow. These are therefore made darker in inking-in the drawing, as exemplified in

EXAMPLE 44, fig. 47, which is the outline drawing of 'front elevation of high-pressure steam-engine,' the plan of sole-plate of which is given in

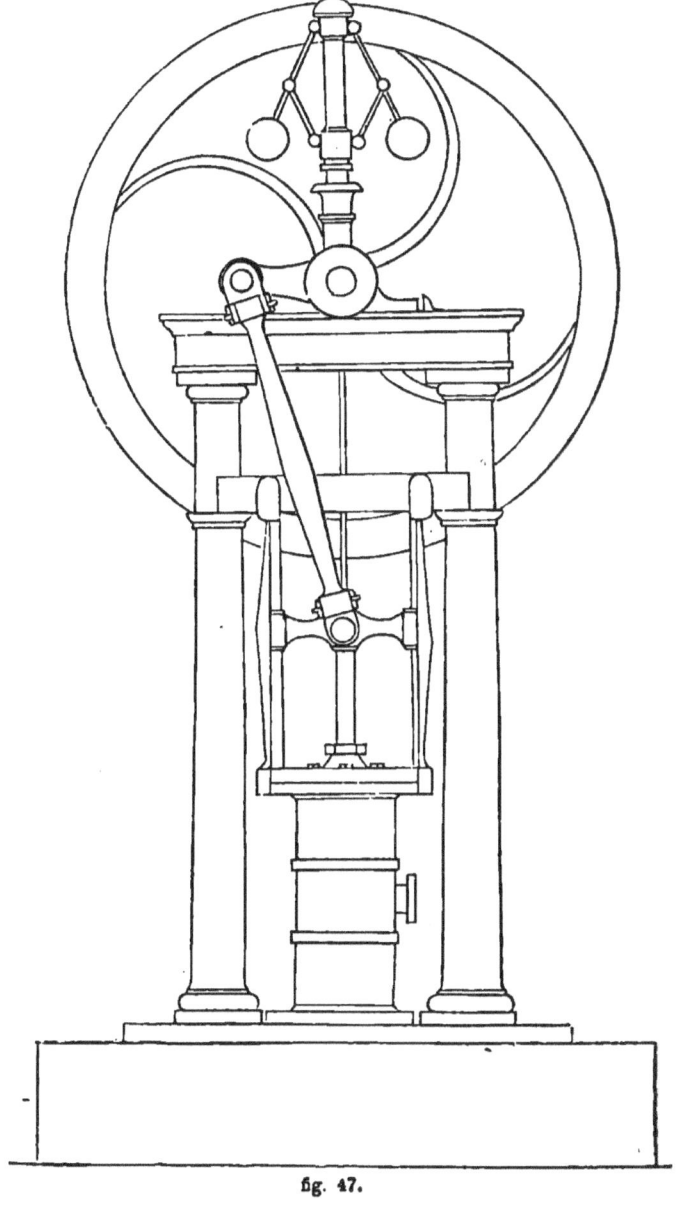

fig. 47.

EXAMPLE 45, fig. 48.

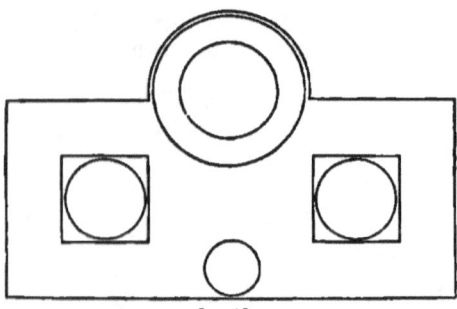

fig. 48.

We now proceed, as a conclusion to this department, to give a few examples to serve as copies to the student, in copying which he will find his operations much facilitated if he has paid full attention to the preliminary lessons. The copies given in perspective are set out by the rule given in the section on 'Perspective' in the *Illustrated Drawing-Book*, to which we refer the reader.

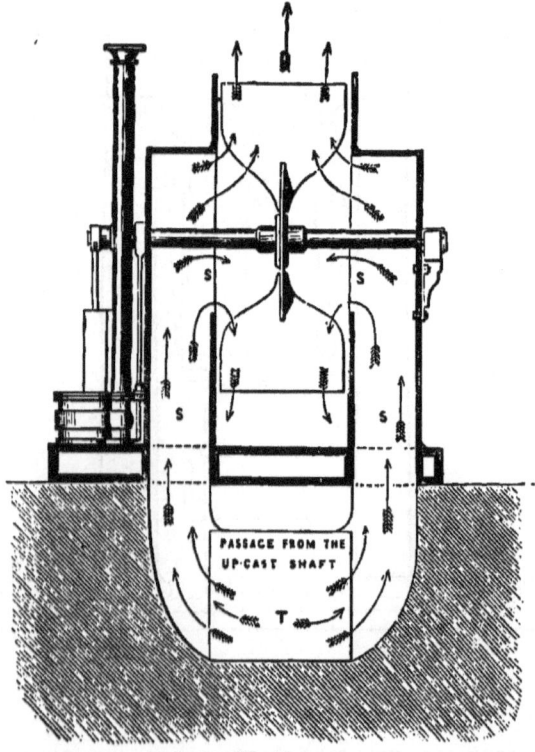

fig. 49.

EXAMPLE 46, fig. 49, is a transverse vertical section of Nasmyth's steam ventilating-fan.

EXAMPLE 47, fig. 50, is a longitudinal vertical section of an aerated water-machine.

fig. 50.

EXAMPLE 48, fig. 51, is a longitudinal and transverse vertical section of a smoke-burning furnace.

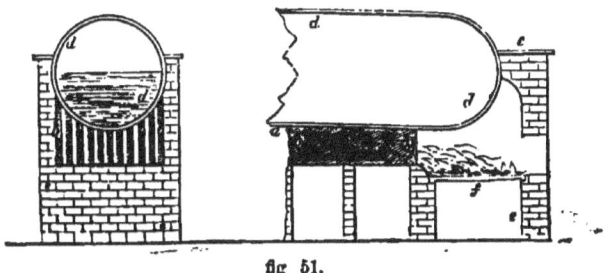

fig 51.

EXAMPLE 49, fig. 52, is 'side elevation' and 'end elevation' of Roberts' Alpha clock.

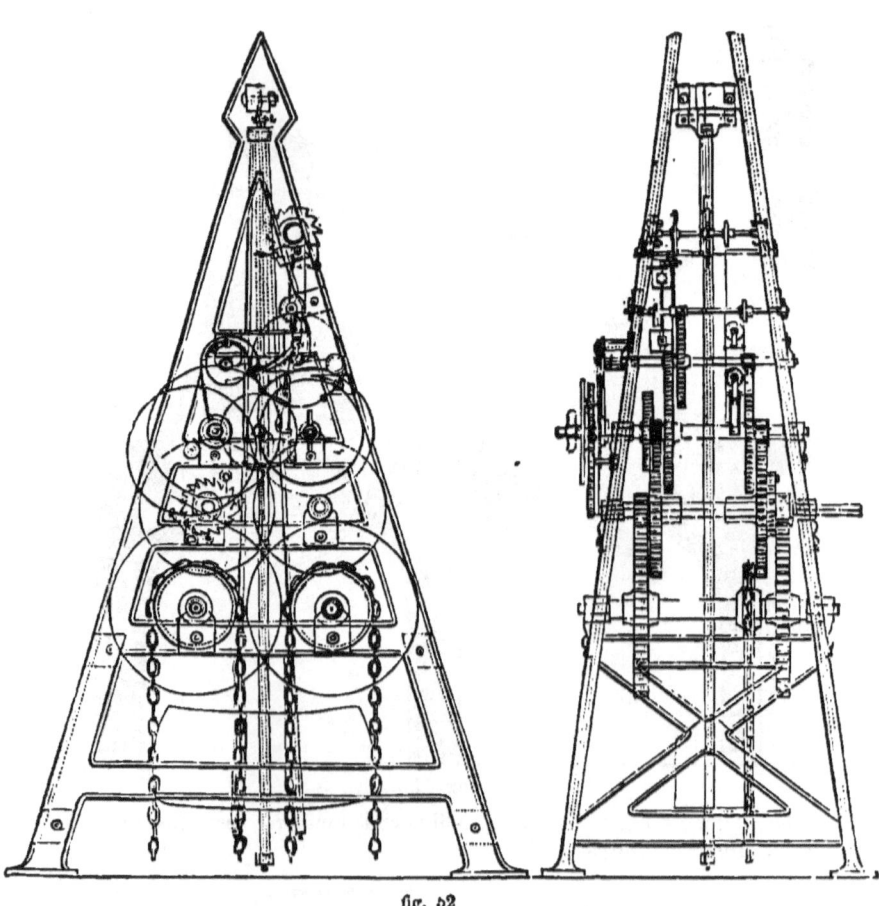

fig. 52

EXAMPLE 50, fig. 53, represents a side-elevation of a corn-mill, with section (vertical) through the grinding-plates.

EXAMPLE 51, fig. 54, is a perspective view of another form of portable corn-mill.

136   ILLUSTRATED ARCHITECTURAL,

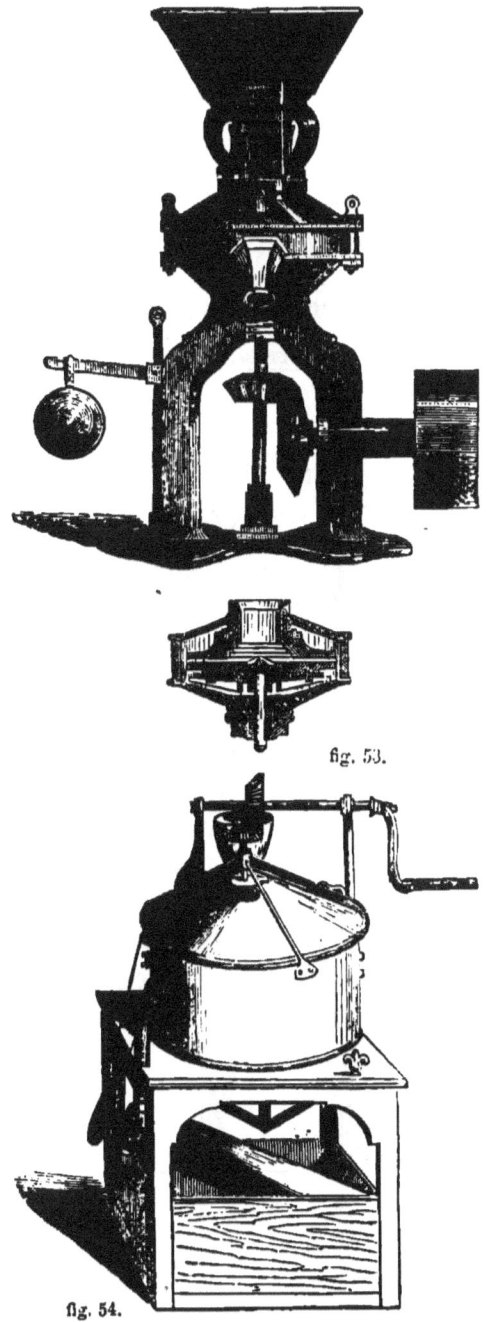

fig. 53.

fig. 54.

ENGINEERING, AND MECHANICAL DRAWING-BOOK. 137

EXAMPLE 52, fig. 55, is a transverse vertical section of the 'patent conical flour-mill,' of which the perspective view is given in

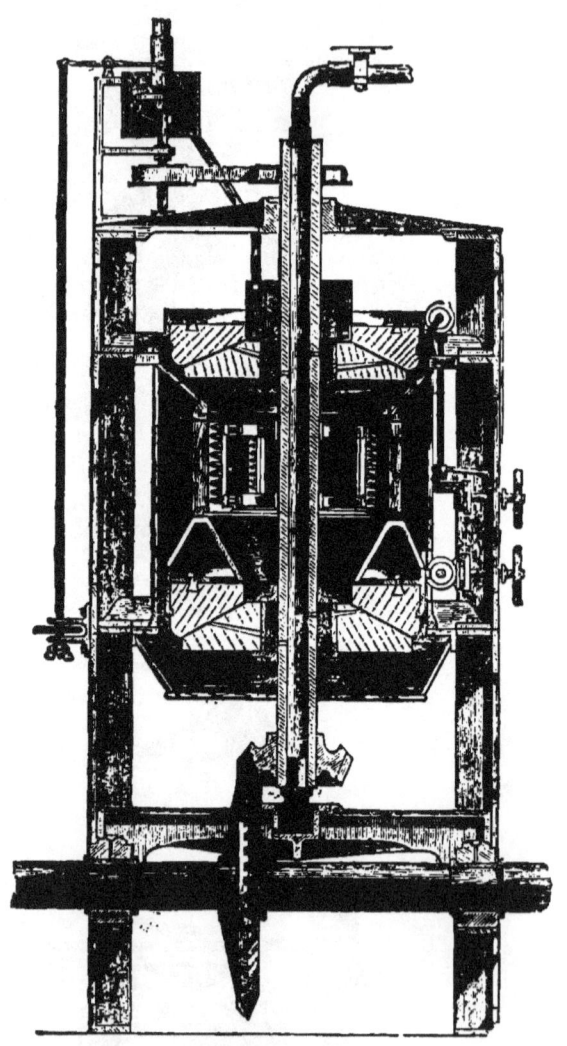

fig. 55.

138   ILLUSTRATED ARCHITECTURAL,

EXAMPLE 53, fig. 56.

fig. 56.

EXAMPLE 54, fig. 57, is front elevation of a fixed high-pressure steam-engine.

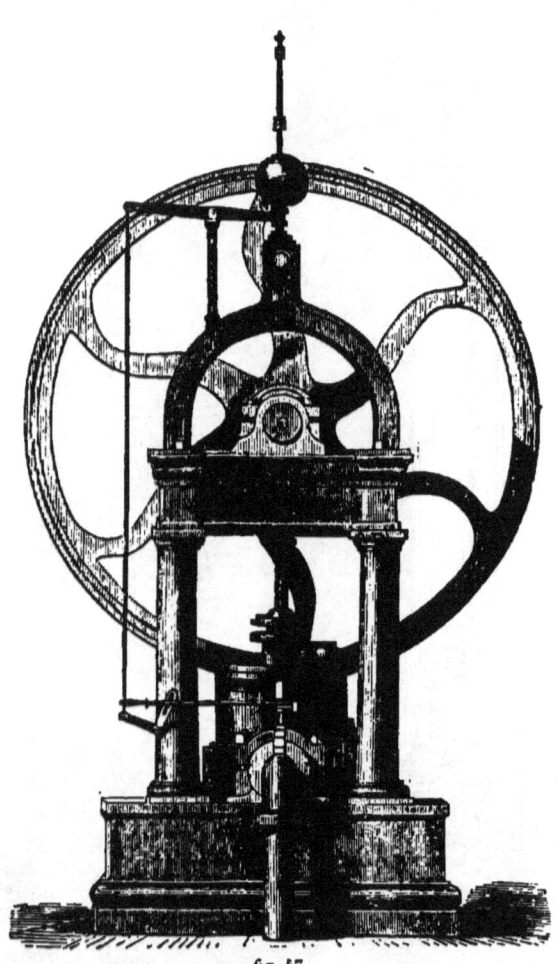

fig. 57.

140   ILLUSTRATED ARCHITECTURAL,

Example 55, fig. 58. is a perspective sketch of a fire-engine.

fig. 58.

ENGINEERING, AND MECHANICAL DRAWING-BOOK. 141

EXAMPLE 56, fig. 59, is a side elevation of a 'disc-pump.'

fig. 59.

142 ILLUSTRATED ARCHITECTURAL.

EXAMPLE 57, fig. 60, is a perspective sketch of a 'drug-grinding-machine.'

fig. 60.

ENGINEERING, AND MECHANICAL DRAWING-BOOK. 143

EXAMPLE 58, fig. 61, is the side elevation of an 'American wood-burning locomotive.'

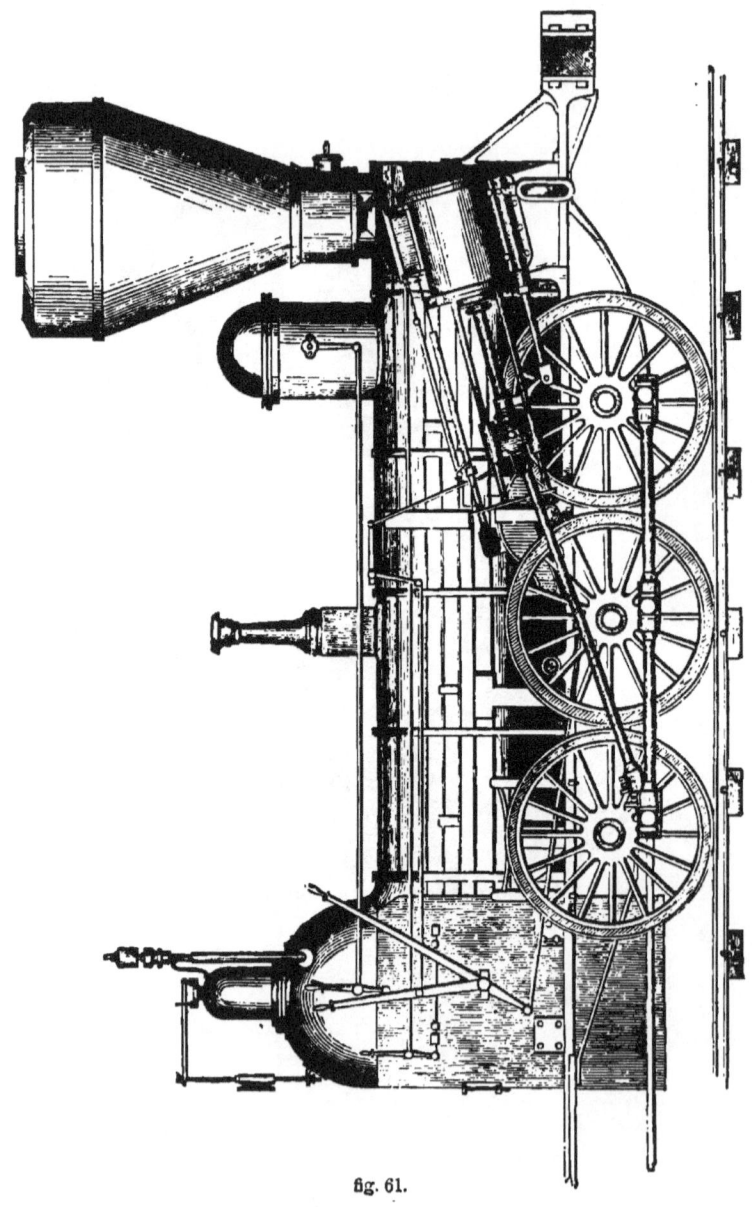

fig. 61.

In the various examples we have given, the pupil will perceive the method in which the various parts are shaded in order to represent round parts, flat, and so on. Mechanical outline-drawings may be shaded by means of lines, as in the examples we have given, thus imitating the manner in which engravers give the desired shade. When this is carefully executed in fine ink lines, regularly drawn, the drawing has a fine effect when finished, accurately presenting the appearance of roundness in some portions, and flatness in others, according as the subject requires. When this method is considered too tedious, the shades may be put in with Indian ink and a camel-hair brush, the appearance of roundness being imparted by first putting in a part of uniform depth in tint, and washing the outside line of this with a brush moistened in pure water, until the colour gradually blends into the tint of the surrounding paper. The depth of tint towards the outside part should be gradually got up to the desired point by repeated operations, the colour used being of a light shade. The addition of a little blue imparts a softness to the Indian ink, which is agreeable to the eye. Cast-iron surfaces are represented by a bluish-gray tint, malleable iron by a light blue; brass surfaces by a faint yellow, brick by a reddish yellow, faintly mottled with a shade darker of the same colour; stones by a faint yellow, with horizontal streaks of a darker tint; wood by yellow, with vertical streaks of a faint black; water by faint blue, with horizontal streaks or lines of a faint black: these look best when put in carefully with the pen and square, as in the diagram in fig. 62. These are the principal shades of colours required in mechanical drawings. The colours generally required are Indian ink, gamboge, Prussian blue, Indian red, lake, and sepia.

fig. 62.

The reader desirous of extending the range of his copies will find numerous excellent examples of machinery in the work on Mechanics in this series.

THE END.

www.ingramcontent.com/pod-product-compliance
Lightning Source LLC
Chambersburg PA
CBHW020057170426
43199CB00009B/311